Seven Ways of Knowing

David Kottler

Hamilton Books
A member of
The Rowman & Littlefield Publishing Group
Lanham · Boulder · New York · Toronto · Plymouth, UK

Hamilton Books
4501 Forbes Boulevard
Suite 200
Lanham, Maryland 20706
Hamilton Books Acquisitions Department (301) 459-3366

Estover Road
Plymouth PL6 7PY
United Kingdom

Printed in the United States of America
British Library Cataloging in Publication Information Available

Library of Congress Control Number: 2010929918
ISBN: 978-0-7618-5189-9 (paperback : alk. paper)
eISBN: 978-0-7618-5190-5

Apologia pro sententia sua.

In fondest memory of the late Dr Simon Nurick who criticised the book in detail and with thanks to Dr Robin Reeves who was also critical about much of the content. My thanks also to my long-suffering family who have been forced to read and proof-read and listen, and especially to my wife Joan for her unfailing support.

CONTENTS

PREAMBLE

This book was started over a decade ago while I was still a practising civil engineer as an attempt to clarify my thoughts of what it means to know and the extent, if any, of sure knowledge.

It starts with an analysis of sense perception which in turn entails some understanding of the nature of space and time. The critical part which evolution has played is stressed; the credence we can give to the theory itself is discussed in a chapter devoted to the vast amount of knowledge we gain not from personal experience but from books or other media.

Whether abstract knowledge and the way we know other persons differ from the way we know material objects are examined as dispassionately as possible, as are ethical and religious knowledge although in all three introspection plays a large part and personal beliefs are not concealed. That these have been influential in the development of the philosophy articulated in this book is undeniable but it is hoped that no dogma has intruded.

Finally there is a theory of perception which offers a possibility of a coherent understanding of all our knowledge.

Two personal crusades against widespread modern cultural attitudes are indulged: that lack of understanding of mathematics is something of which to be proud and that atheistic materialism is virtually a scientific fact.

David Kottler Cogenhoe Northampton UK September 2009

INTRODUCTION

A philosophy may be termed holistic if there are no limits to the field of its enquiry; it may be judged by the extent to which it can inform every conscious action.

By a philosophy is meant a system of thought which after as searching an examination as possible of the assumptions which underlie all sciences, attempts to provide a coherent framework for them. This examination will be an ongoing process which will only be completed when the whole system is coherent.

To inform in this context means to heighten the awareness of any conscious act to enable judgements to be made concerning it.

For present purposes conscious acts are limited to human actions which have an element of self awareness. Eating would be included; digesting the food would by and large be excluded.

The first task of any philosophy is to know what it means to know, to understand what is meant by understanding.

This should not be looked on as a dogmatic statement. Although I have a reasonable idea of where I want to go, I have at the moment only the sketchiest notion of how I am going to get there. A start has to be made somewhere.

In order to begin to know what it means to know it might be useful to enquire into our ways of gaining knowledge.

Assuming that knowledge is possible.

1st Assumption: We have certain knowledge of a huge and diverse range of things.

There is no intention here of assuming what knowledge actually is. One of the main tasks of this work is to see if that the assumption is reasonable.

2nd Assumption: Everything can be subjected to reasonable enquiry.

That is to say, there are no perverse or illogical forces operating in the natural world. This will be examined in greater detail later but for the moment consider the outcomes of simple games of chance.

It is possible to forecast with great accuracy the statistical likelihood of a particular number coming up on a roulette wheel. Indeed, any anomaly will be used as an indication that the wheel is not perfectly balanced. The reason that any particular number comes up one in thirty seven times is that there is no reason why it should come up more often.

1) It might be pre-programmed in the form of genetic information.

Knowing what to do instinctively is clearly a form of knowledge.

2) It might be innate and immaterial.

As for example inborn knowledge of good and evil.

3) It might be gained through extra-sensory perception

For example, divine inspiration, telepathy, sortilege.

4) Finally, it might be gained through sense perception.

Sense perception includes inner senses such as headaches, feelings of fear or elation and the like.

For the moment it is only this latter form of knowledge that will be investigated.

We can only discuss, talk about, think about, matters of which we have some knowledge of some sort. This is obvious although there are cases which are the subject of much thought and controversy and where the content of the thoughts is (for most of us at least) pretty hazy, such as the concept of God.

A word has to have some significance if it is to be the subject of rational discourse: we have to have some knowledge of its meaning. I am not at the moment concerned with trying to categorise the various sorts of knowledge; for now let us consider some arbitrary types of knowledge, or things that we know, or things that we wish to discuss.

I know a tree

I see it daily, touch it frequently, hear it when the wind blows: it is a good example of any material object in the ordinary world we inhabit.

I know there is space between us and the moon

But what space is I have only the vaguest notion

I know that two and two make four

I might know it by rote; I might understand it

I know that Athens is the capital of Greece

I have been told so and have read it in an atlas

I know Mr Smithi

I know him to say, 'Good morning'.

I know him intimately; he is one of my best friends. In the former he is being treated more or less like a tree. In the latter has something quite different crept in?

I know that I sometimes do wrong and am reasonably sure that when I am absolutely honest with myself I know the difference between right and wrong for me without having to think about it

Although it is possible that this might be innate knowledge it is very often the product of indoctrination and although the truth of the general statement is obvious it is also true that we frequently feel guilty of things over which we have no control and proud of others of which we should be ashamed. Our knowledge is at best uncertain and at worst plain wrong.[1]

Adam knew Eve

As far as I know, a use restricted to the Bible. Carnal knowledge is outside the scope of this work.

I know that my Redeemer liveth

St Paul said this, quoting Job. He had an experience which I have not had. I can only say, 'I believe that my Redeemer liveth,' even though I would stake my life on it.

Each of these forms of knowledge will be the subject of a chapter, in the order in which they are listed above, but under the more formal titles and with an introductory chapter and a concluding one:

1) The basis of perception and knowing
2) The perception of space, time and matter
3) Abstract knowledge
4) Book knowledge
5) Knowing persons
6) Morality: the knowledge of good and evil
7) Religious belief
8) A theory of perception

1. Here (and this is an aside) I come to an ambivalence I have never resolved. There is the opinion written opposite and the thought that it would not be reasonable not to use every fibre of our intelligence in trying to come to any judgement and particularly a moral one.

CHAPTER 1

THE BASIS OF PERCEPTION AND KNOWING

There once was a man who said, "God . . . .."

There are, among others, two trees in our garden which we (my wife and I and, for as long as they lived here, our children) have seen most days for the best part of forty years. One is a large walnut which when the wind blows we can hear sighing or recognise its creaking, although unlike Giles Winterborne we would not have known it was a walnut just from its sound. We have felt it and clambered over it, swung from it and built tree houses in it over the years and even put it to ignominious use as a clothes-line prop. We and countless mice, squirrels and various birds have often tasted the fruit which it occasionally bears in great profusion. It has not got much smell. The other tree is a berberis which in May smells just like a well known brand of lavatory cleaner.

When I see them, it is believed that light radiating from the sun is reflected from their leaves and bark in continuous streams of unimaginably large numbers of incredibly tiny rather mysterious entities called photons.[1] These enter my eyes and activate some of the millions of photo-electric cells in the retina upon which they have been focused by the lens just behind the pupil. By a complex chemical transformation pulses of electricity are transmitted along the optic nerve into the brain where after several million electro-chemical interchanges all on a microscopic level and a small but measurable interval of time I see a tree. Similarly when I feel a tree, pressure sensitive transducer nerve endings in my skin transmit minute pulses of electricity along the nerves in my arm, up those in the spinal cord and finally into the brain where again after a very small but measurable time lapse I experience the sensation of roughness of bark, or the smooth immensity of a large branch. Similar processes occur with the other types of sense perception: hearing, tasting and smelling. In each case an external stimulus is converted to a very complex system of electrical pulses in the central nervous

1. An examination of the current state of knowledge or belief of the nature of photons follows later. For the moment let if suffice that as far as we know all forms of electromagnetic radiation including heat and light consist of photons which exhibit characteristics both of waves and particles; they can be absorbed or emitted by matter, and that all physical, chemical and biological reactions are the result of such emissions or absorptions.

system which results in our having a sensation. The sensation is as remote from the object sensed as you are from some one distant of whom you have only a photograph. When we see a photograph of some one, we know we are only looking at a photograph and not at the person himself. When we look directly at some one we merely get a better photo: more immediate and probably in greater detail but still a sort of photo. It is somewhat analogous to looking at some one through the screen of a digital camera. What we experience is a representation, some sort of coding, of the person seen. However complete the coding might be, it is still a coding and not a person. We have no direct contact. Between the object and the perception of it, its image is successively encoded, stored and transmitted, many thousands of times and in widely disparate forms. The final product, and one that might or might not be of an entirely different order from the rest, is the conscious perception itself.

The actual sensation is manufactured within the central nervous system. Everything that we see or hear or smell or touch or taste is contained within our heads by the time that we see or hear or smell or touch or taste it. We have no direct contact with anything we physically experience.

It is a commonplace that when we see a distant star its light has taken a thousand years to reach us, and therefore it is not the present star we see but one a thousand years ago but this is not the point I am trying to make: I am perfectly happy with the notion that 'now' is best defined by the passage of light (of which more later). Whether its light has taken a microsecond to reach us, or a thousand years, we are remote from the object we see.

Just as the screen on a digital camera has to be constructed to produce an image as close as possible to the one that we experience but could easily be made to show all reds as greens, or all straight horizontal lines as wavy oblique ones, so it can easily be imagined that quite a small variation in the light cones in one's retina would result in seeing a quite differently coloured tree. As it is I do not see identically out of each eye. My wife and children each have at very least slightly different pictures and possibly very different ones: none of us has access to anybody else's consciousness.

A few crossed nerves and a different cerebral system and perhaps I would not see the tree at all. I might hear it. A stone or a mountain would be a monotone in stable light and turn to music only with the passing of a cloud. Imagine the harmony of a silver birch in a spring breeze or the ripple of long grass in the autumn.

It might not be imaginable but it is certainly not unthinkable that given a completely different nervous system our perception of a tree would be completely different, *completely, absolutely and wholly different*; for example the tactile experience of putting our arms round its trunk could be perceived as a colour. Dwell on this thought for a moment or two—or a year or two (or for me forty years because I am slow witted). As I let my imagination wander, so my perception of substance dissolves and fragments.

If evolutionary theory is to be believed (and it is certainly not being questioned here) the nature of our perceptions is the result of the accumulation of millions of tiny chance mutations over aeons which, for whatever reason, have endured. If this is the case it would seem that the whole of our sensory apparatus and the entirety of our sensations, are an accident of evolution. That there is a teleological element (that evolution is directed to some end) is rightly anathema to evolutionists: the thought that we have evolved in order to see the world as it really is, is clearly absurd. It is an incontrovertible fact that there are many energy changes between an object and our sensual perception of it and that the sensation itself is manufactured within our central nervous system.

This is not denying for a moment that the world around us exists independently of us: it is not a descent into solipsism. It is saying that our perception of the world is conditioned by the means by which the perception is gained and that we have only an indirect access to it.

Our every sensual experience is centred on ourselves set in a three dimensional space at a particular moment in time that creeps or leaps inexorably from past to future. As I dwell on the indisputable, the incontrovertible fact that all my sensations, my perceptions, are constructed within my nervous system, my thoughts of space and time and matter themselves become hideously confused. What we know of them is the subject of the next chapter. Some comprehension of the underlying physical structure of the universe is necessary if we are to understand the nature of our perception of it, and so a chapter is devoted to an explanation in plain language of the theories of Relativity and Quantum Mechanics. A fuller and more technical treatment is included in an appendix.

The remainder of the book can be read without recourse to this chapter, but it would be a pity!

Can we begin to know the nature of reality or to begin to understand what things are in themselves? Can we really know anything? Are these real questions or is their content void? For the moment I can only explore, only speculate.

With the possible exceptions of religious faith and innate conscience all forms of knowledge have as their ground sense perception, but the connections between a sense stimulus and the perception, and between the perception and knowledge remain profound mysteries.

Philosophy and science have jointly tackled this problem since the Greeks began their great investigation. Religions have had their say. Newton appeared to simplify at least part of the make-up of the universe but a critical analysis uncritically using some of his basic assumptions led to the antinomies of Kant.

Where then does reality lie? Is it in the things themselves, the tree, the solar system, the universe which we can only experience, as it were, at a distance, and through varying transforming media? Do we see an object more really when we examine it through a microscope or an electron microscope, and learn more and more about its constituent parts? Are we closer to the reality of a watch when we understand exactly how it works, or to the nature of the solar system when we comprehend the simplicity of a Copernican rather than a Ptolemaic system? Is there any sense in thinking that the sun does not really rise and set, that this is merely an illusion caused by our positions on a planet which really encircles it?

Despite these thoughts, one of the recurrent themes of this work is that there is nothing more real in the material universe than our conscious perceptions of it. Following the spectacular success of Newton in the seventeenth century it became engrained in western consciousness that we live in, and all material objects exist in, an absolute time and space; we can observe this and get to know it dispassionately and this is how most of us still think. The acute observations of Berkeley and the doubts that he expressed concerning the Newtonian, (or what has become 'common sense') view of material existence appeared fantastical at the time and we have great sympathy with Dr Johnson's expostulation as he kicked a large stone, "I refute him thus". Nearly three hundred years of sceptical philosophy, from Hume onwards and the three revolutions in scientific thinking in the nineteenth and twentieth centuries, Evolution, Relativity and Quantum Mechanics, have done little to alter this perception even though they absolutely contradict it. Even among religious believers the intervention of a deity in the creation and running of the universe has been lessened step by step as various mysteries have been explained: the evolution of species, the equivalence of mass and energy, the causes of earthquakes, the cure of disease, the nature of madness. Yet it will be suggested in the final chapter of this book that the belief in the existence of God could lead to a glimpse of at least one possibility of an altogether different, fuller and wholly satisfactory understanding.

This is not to be taken to denigrate in the least science or scientific research. Of course we must strive to understand how things work and, such is the breadth and complexity of scientific knowledge, we might usefully and fulfillingly spend a lifetime advancing a tiny specialty. We will not thereby get any closer to knowing what things in themselves are; we will get a deeper understanding of why the sun rises in the morning, why a vortex forms when the bath plug is removed and why eating a sour apple gives us a stomach ache.

Great advances in the knowledge of electromagnetic radiation and its implications on the nature of matter were made in the nineteenth century and it looked for a time as if progress was being made into the understanding of the real nature of the make-up of matter. During that century Darwin's theory of evolution began to radically alter our perception of man's place in the order of things, from being observers standing at a distance as it were from the rest of creation. It was still assumed that three dimensional space was a homogeneous entity in which we dwelt and that time ticked on at the same unalterable pace from the past into the future at all places and all times. Then the two great scientific theories of the twentieth century, Relativity and Quantum Mechanics, threw into doubt such notions; they operate on the very edges of our ability to understand so that after more than sixty years of intense questioning there is still no consensus as to what the theories "mean". At the same time as relativity and quantum theory were being developed, philosophers were tackling the age-old problems largely through an examination of logic and language. Although this mass of writing has served to make one much more critical of the use of language it has at least in my own case been largely sterile and its results are summed up in the prefaces to Wittgenstein's two great books, written either side of A J Ayer's seminal Language Truth and Logic. In 1918 Wittgenstein wrote in the preface to the Tractatus, "On the other hand the *truth* of the thoughts that are here set forth seem to me unassailable and definitive. I therefore believe myself to have found, on all essential points, the final solution to the problems." The truth that he had found was that when an answer could not be clearly articulated, neither could the question: "Das Rätsel gibt es nicht" (The riddle does not exist)[2]. Almost thirty years later, and twelve years after Ayer's Language Truth and Logic, in the preface to his Philosophical Investigations he said he had been forced to recognize great mistakes in what he had written in the earlier work.

Each major branch of science has thousands of sub-branches and in many of these there is a lifetime's study. Is any meaningful overall understanding possible, or because of the huge expansion of specialist knowledge is it just impossible to have any real understanding of some or other fundamental question? For instance, Richard Dawkins in The Blind Watchmaker says that the understanding of evolution is his business but that he can accept on trust a highly controversial and self-confessedly wholly speculative theory of the creation of matter.[3] An understanding of evolution is relatively easily gained by reading a book such as The Blind Watchmaker (or perhaps more accurately an understanding of one and a critique of many other theories of evolution) but to be able to understand a theory of the spontaneous coming to be of the universe such as Peter Atkins speculates requires some real understanding of both quantum theory and relativity and each of these in turn needs some maths. Nothing can replace personal understanding. A student of a particular science may perhaps legitimately confine himself to that science; a philosopher may not. That sounds suitably pompous. If some maths is needed, then it will have to be learnt. After all, anybody who can tell the difference between a logical and an illogical argument, and that includes pretty well all of us (though some are quicker than others, but that is not important) can easily be brought to a real understanding of sufficient maths to be able to grasp the basics of Quantum Mechanics and the Special Theory of Relativity. A brief summary of the necessary mathematics, assuming all previous knowledge has been forgotten, forms the subject of the first appendix and its application to the two theories is dealt with in the second. The General Theory of Relativity and refinements of Quantum Theory are expressed in much more complex maths which require a real effort to understand and it is questionable whether the increase in overall understanding merits the necessary effort. With fairly simple maths it is possible to begin to see how these theories fit into a general understanding of all that there is to understand, at least of the physical relationships of the universe.

So far I have touched (that is an interesting use of the word) only on the perception of things seen, but exactly the same goes for all the other senses. There are a number of energy transformations between the striking of the drum

2. Tractatus 6.5

3. The Blind Watchmaker, Penguin 1988 Page 8; the Creation, Peter Atkins, Freeman 1981 page 115.

and the throb, the removal of the stopper from the perfume bottle and the scent, the prick of the needle and the pain, the popping of the chocolate into the mouth and the delight (or sometimes, especially as we get older and our gums leave the bases of our teeth exposed, excruciating pain).

In addition to the external senses there are of course internal aches, pains and pleasures which in the same way are mediated by a number of energy transformations between the cause and the sensation. Even in extrasensory perceptions such as telepathy or mystical experiences there might be some cerebral activity but they are outside the scope of at least this chapter in so far as there is no satisfactory scientific evidence of their existence. (This is not the same as saying that they do not exist.)

Both the Oxford English Dictionary and Webster's give perception as a primary meaning of *know*; neither mention knowing among the various meanings of perceive.

As we perceive the material world, so its images enter our memory, and once there we say we know the things which gave rise to the images. Very little is known at the moment about the mechanics of memory although theories are beginning to emerge. They do not concern this book except in so far as they confirm that permanent or semi-permanent physical modifications of brain cells occur and that we do access them when we remember. The manner in which these cells are available to those parts of the brain responsible for perception is still a complete mystery. When we do remember something we are, I think, in essence perceiving a memory: we are not perceiving a perception. The memory of things, even of abstract things, is a form of sense perception, in the sense that it is an operation that requires the activity of the central nervous system but in remembering something, we are not perceiving a perception: we are perceiving something within us in essentially the same sort of way as we have a headache or, which is much the same thing, become aware that we have a headache.

Things which we cannot see or hear or touch or taste or smell, such as truth or maths or love form the subject of Chapter 3 in which the primary subject is the nature of insight. What is the difference, if any, in my saying that I know a particular tree and that I know what a tree is and that I know that two and two make four? I believe that in knowing a particular tree or a dog or a rainbow, the only difference is in the object known: the knowing is of the same nature and the word *know* could be replaced by *can recall from my memory* in all instances and still remain accurate. That the word *know* could be replaced by *can recall from my memory* is not so straightforward when I say that I know that two and two make four, unless all that is being claimed is a parrot knowledge without understanding. No element of insight is needed in knowing a particular tree but what about the difference between trees and giant ferns?

When I see a robin I recognise it as such. No conscious thought process occurs; the recognition is as instantaneous as a bulb lighting when the lights are switched on. In both cases a short time interval does of course occur. Sometimes I see a bird which I recognise but I cannot think of its name (actually, for me this happens more frequently with flowers than birds) and no amount of trying to remember seems to help in the least except in this respect: that it is not uncommon if I really try then the name does come back to me several hours later. In each case more than one item is being recalled from memory: a former robin, or several former robins or a picture of a robin and an associated name; the recalling from memory is partly spontaneous and partly willed—and partly unsuccessful. Our access to our memory banks is very uncertain.

My car radio has a search function and in places where reception is poor and I want to change the wavelength so that the reception is improved I push the appropriate button and notice the digital read-out going round and round in a circle, starting at eighty megahertz, going rapidly through to a thousand and forty and then starting again at eighty, over and over again until at last it latches on to some wretched local station playing a cacophony so I start it all over again and finally I find the programme I am looking for. I wonder if something vaguely similar is going on in the brain when we try to remember something or to understand something—continual activity with finally a connection. When I hear a piece of music which I do not recognise, I play a game and say to myself, "It is clearly a little later than baroque, classical rather than romantic, it has a clarity that is only found in very few composers, say Hayden or Mozart or Beethoven and it has echoes of each of these but something does not quite fit. How about Schubert? Now there's a thought. . . ." When at the end the announcer says "That was Schubert's trio in B Flat" I get a tingle of satisfaction. (When he says, "That was the second movement of Shostakovitch's piano trio", I think I actually blush and thank my lucky stars that no one else was privy to my mental process.) Knowing a piece of music is of the same nature as knowing a tree or knowing Mr Smith: the music is a material object and the act of knowing is one of conscious sense perception. Forming a right judgement about the composer, recalling where I have seen Mr Smith before, are these a different type of knowing? We cannot touch or taste or smell a judgement—and yet we say that we feel the truth, that a judgement stinks, that we see the logic or, in today's cant, "I hear what you say".

The mechanism for the perception of parts of our bodies such as our fingers is exactly the same as that for perceiving any other material object: If I close my eyes and interlink my fingers pressure transducers in the skin convert touch into tiny electrical impulses which are transmitted to the brain; my fingers (language here slips into inadequacy) become conscious. If I am playing a piano *I* do not send messages down to *my* fingers; rather, it is the fingers that are making the judgements. If I close my eyes and try to imagine my fingers, I can form some sort of mental image of them (in my dreams I can see them or something very close to seeing, in great detail and clarity). What I am interested in at the moment is the essential difference, if any, between my perception of my fingers when I am looking at them or feeling them and my perception of them when I am doing neither of these things but relying on memory. Now, I can sit for ages with my fingers interlinked while I am talking to you or while I am working something out, and be quite unaware of them and they can be out of sight and sound and touch and yet I can be acutely aware of them. (Particularly when I am keeping my hands behind my back in the hope that no one will notice my filthy nails.) In both cases there is reciprocal nervous activity. In the first case, neuron activity in the finger tips triggers neuron activity within the brain which results in my fingers feeling. In the second case neuron activity within the memory cells sets off sets off activity within that part of the brain which evokes consciousness and leads to a stream of memories.[4] In both cases (the touch sensitive neurons in my finger tips and the memory cells) are themselves switched on or otherwise controlled, sometimes voluntarily or consciously and sometimes involuntarily or consciously and sometimes involuntarily or unconsciously, by another sort of neuron activity within the brain. In both cases there is a complex chemistry of the whole me involved and not some part of me observing as it were some other part. The essential difference would seem to be that consciousness arises out of the action of different banks of neurons within the brain but in each and every case it is the same whole single me who is conscious. Our knowledge of the interconnectivity of the central nervous system is still extremely sketchy but I think that the form of mechanism outlined above is broadly true. If so, there does seem be a possibility of arriving at a vestigial understanding of the relationship between being aware of material things and of remembering them and the ability to compare and combine both of these and to abstract from them. The key points that I am trying to articulate are that all consciousness arises from neuron activity within the brain and that this activity may be turned on by impulses coming from various parts of the body including that permanent or semi-permanent data base within the brain itself which is the physical ground of memory. It is this ability to be aware of stored information which makes comparison possible: not from the perception of a perception but through the perception of the memories of perceptions.

This distinction, which might appear to be a quibble, a minor technical point, is crucial to an understanding of this theory of knowledge. Let us trace a

---

4. This thought, that there is some area in the brain whose function is to awake consciousness, is strongly challenged by Daniel Dennett in the following passage from Freedom Evolves, Allen Lane 2003, Chapter 8: "Mental contents become conscious not by entering some special chamber in the brain, not by being transduced into some privileged and mysterious medium, but by winning the competitions against other mental contents for domination in the control of behaviour, and hence for achieving long-lasting effects–or, as we misleadingly say, "entering into memory". To a supposed objection: "suppose that, as you say, the conscious processes are simply those that win the cometitions. How does *that* make them conscious? What happens next to them that makes it true that *I* know about them? For after all, it is *my* consciousness, what I know from the first-person point of view, that needs explaining!" Dennett answers this: "Such a question betrays a deep confusion, for it presupposes that what *you* are is something *else*, some Cartesian *res cogitans* in addition to all this brain-and-body activity. What you are just *is* this organization of all the competitive activity between a host of competences that your body has developed."

It seems to me that had the objection just left it at "How does *that* make them conscious?" the mystery still remains and the Cartesian red herring avoided!

simple (simple??!) sense perception. A single photon from a minute flash is absorbed by one of the hundred and twenty-five million photocells in the retina of a human eye and there is a discharge of an electric current—a movement of electrons—by a complex relay system along the optic nerve into the brain. Here, at least two separate but connected things happen. In one part of the brain some activity involving electrons and photons, triggered by the discharge of electricity down the optic nerve lays down a permanent or semi-permanent record of the capture of the original photon in, for want of argument, a single memory cell. At the same time (or more accurately after an interval measured in tenths rather than thousands of a second) and probably but not necessarily in another part of the brain which awakens consciousness, some activity occurs also involving electrons and photons and I see a scintilla of light. The time delay seems to vary between one tenth of a second and a second, depending on the nature and continuity of the stimuli: some editing and even reordering takes place. Of the form of these cells we have at the moment almost no understanding. The memory cells can at a later time discharge currents into the consciousness-making part of the brain and when this happens we will have a memory of the scintilla. The activity of the memory cell is in its turn triggered by an incoming impulse which can originate from various parts of the brain for various reasons. . . . The sheer scale of the complexity of the interconnections is mind boggling. (What I mean to say is that I cannot get my head round it but then in reality that is precisely what my head is.)

What is certain is that the conscious act is created in the brain, and that it is the whole me who becomes conscious.

Sense perceptions can be routed through memory cells to motor cells without impinging on consciousness. A common instance of this is the well known phenomenon of blind sight. It is quite possible, for example, to drive for a hundred miles while engaged in an absorbing conversation or day dreaming, and have no knowledge of where one is when consciousness of the road returns, and no recollection of where one has been. Yet all this time all the complex reactions to roads, traffic and signals have been taking place.

How much of things of which we are conscious are stored in our memories, though we cannot recall them at will? Are sense perceptions of which we are not conscious ever stored? I would guess that the answers to these are a high proportion of the former and a lesser fraction of the latter, but this is largely speculation.

If, to pluck figures out of the air and move for a moment into pure speculation, we had a million sense receptors firing at a rate of ten times per second for sixteen hours a day for eighty years and each one was to leave a permanent memory then something of the order of twenty thousand million million locations (a twenty million gigabyte RAM) would be needed. The entire brain consists of perhaps one hundred thousand million huge specialised cells known as neurons so if half of these were to form the memory, each memory neuron would need to have four hundred thousand locations. Specialised cells with this capability are not unthinkable. As they become filled by memories they would be more difficult to locate and would fail more often. Perhaps this is part of an explanation of why the elderly sometimes have excellent long-term memories but no recollection at all of a visit paid earlier in the day. It is known that short-term memory and long-term memory have different locations within the brain and the interaction between these might lessen the need for anything of the order of twenty million gigabytes or it might be that numbers far in excess of this are actually in operation.

Just as a computer can, by simple logical circuits and electromechanical processes, compare inputs of data and sort them into smaller, equal and larger, or into alphabetical order or work through chains of syllogisms or whatever, so, of course, can that super-computer, the brain. (This will, I would say it goes without saying, use biological circuits which can in part be simulated by a computer and not the mechanisms that a digital electronic computer uses.[5]) The internal clocks which run the switching mechanisms on which digital computers operate, tick, at rates in the order of one thousand million cycles per second (1000 MHz or 1 GHz). These are still some way from catching the brain whose internal operating speed is probably of the order of a few hundred times faster but they will no doubt catch up over the next however many years. Their memory capability is also increasing exponentially and might in a very large computer already equal or even surpass that of a human brain.

It might be instructive to look at how a computer's memory works and how it can compare memories to try to come to some understanding of how by transferring electronic digital mechanisms to biological ones it is possible to conceive living memory systems. Similarly a study of an electronic conditioned reflex can provide a vague insight into the way that living conditioned reflexes occur. These topics require an understanding of transistors and logic gates if they are to be of any value at all but for the moment only a very brief elementary schoolbook description of a simple comparing circuit will be given. Electronic circuits, known as logic gates, consisting of two main components, transistors and resistors, can be wired so that various combinations of electrical inputs can result in desired outputs. A transistor is typically a small sophisticated crystal of silicon which under the right conditions allows an electric current to flow through it in one direction only. A resistor restricts an electric current passing in either direction.

---

5. In a joint project between IBM and the Ecole Polytechnique Fédérale de Lausanne it was announced on 6th June 2005 that a simulation is to be built of a neocortical column of a rat, a cylindrical element approximately one third of a millimetre in diameter and three millimetres long, containing some 10,000 nerve cells. The project is expected to take two to three years. There are about one million columns in the human brain each with some 60,000 cells. (Economist Technical Quarterly Review 11th-17th June 2005.) See update in The Guardian, December 20th 2007 reporting the completion of the first phase in December 2006.

If a circuit is now wired with two input terminals and one output so that no current flows out of the gate when the inputs differ but only flows when they are the same a simple comparing cell can be constructed.

Input1	Input2	Output
On	On	On
On	Off	Off
Off	On	Off
Off	Off	On

This particular circuit is known as an Exclusive Nor (or ENOR) gate. If a current flows from the output terminal then we know that the inputs must have been the same.

Many millions of these circuits with complex interconnections can be etched into small single silicon wafers (chips) allowing thousands of inputs and outputs to be combined, all controlled and synchronised by clocks, themselves chips or parts of chips, operating at half a million cycles per second. All sensation and all thought is also mediated through the activity of neurons which either fire or do not fire: which are either on or off.

Physiological perception differs from electro-mechanical processes not only in its biological mechanisms but in its ceaseless self-renewing exploratory activity. Just as the very fabric of all living tissue is being constantly changed, so our perceptions, our thoughts and our memories are ongoing unimaginably complex processes. A simple illustration of this is seen in Wittgenstein's Duck-Rabbit[6], a line drawing that switches not quite predictably from rabbit to duck and back again as you look at it. Neither abstraction nor abstract knowledge is therefore necessarily involved in acts of comparison. What is the difference between saying "I know a tree" and "I know that two and two make four" if a computer can carry out each of these operations, which it obviously can.

First of all, a computer can only carry out any operations because it has been programmed by an external intelligence to do so: it cannot have insights. I am almost persuaded, via Penrose-Gödel,[7] that there are mathematical insights which cannot in principle be simulated by any computer. My train of thought, which I find I am reluctant to make explicit, is that if a computer can be constructed to simulate abstract thought, then there is nothing mysterious in a brain being able to perform abstract operations and there is nothing essentially different between knowing a tree and knowing that two and two make four.

Is it sufficient to say that a computer has no insight into the fact that two and two make four but only that it has been programmed and the fact is within its memory? Is it similarly sufficient to say that a bird's brain has been genetically programmed, because this is the result of evolution much as the human brain is. A bird does not know that two and two make four. It knows things like

6. Philosophical Investigations IIxi p194 Blackwell 1963
7. Shadows of the Mind, Chapter 2. Roger Penrose Oxford 1994

the way back to its nest and how to catch worms. It might just know that this worm is larger or juicier than that but none of these things are the result of insight. Chimpanzees and children can be trained to recognize that two and two make four and it merely becomes a matter of memory. Even at the age of seventeen I managed to memorize sufficient facts of differential and integral calculus to be able to pass A Level maths without understanding: it was not until I had to re-work it all out some years later that the insights came and I really understood what it was all about. Did this act of understanding, this insight, require something of a different nature altogether?

Take the learning to read. The recognition of the association between the symbol **a** and the sound similarity between bat, cat, fat, hat, mat, pat, rat and sat is generally a matter of conditioned reflex and it is this and not insight which will prompt a child to be able to read vat and brat as they have sounds which are (damn) the same (double damn). A child can say, "I know what those letters spell", and he can say, "I know Auntie Mabel". I see no reason why these two operations (knowing what letters spell and knowing Auntie Mabel) could not be programmed into a computer or could not have evolved in humans in any way separately from the way any other animal has evolved. Neither, it seems to me involves abstract thought.

If you teach a child to recognize an elephant and then ask her why she thinks that a particular animal or picture of an animal is an elephant there are a number of possible responses. She might say, "Because it is", or she might say, "Because it has got a trunk". She might search her memory (unconsciously) in a way analogous to that in which we as adults work our way through a flora when trying to identify a wild flower. Even when we have correctly identified the flower I do not think we will have had any insight into the matter, and once again we will have done nothing that a well-programmed computer could not have done.

Knowledge not gained from personal experience but gleaned from reading, or listening is treated in Chapter 4 with particular emphasis on its reliability. Physical theories, such as gravity, evolution and global warming are examined not in themselves but in the credence we can give them. It is pragmatic rather than philosophical and might come as a welcome relief from the more demanding chapters that precede and follow it.

Chapter 5 looks at the way we know people, including ourselves, for it is possible that something quite outside the remit of physics is necessary, that an immaterial element is present.

I remember when I was very small, at infant school, so less than eight, I was bored with a lesson and stuck my hand up and said, "Please may I be excused" and went to have a wee. I did not really want to go, and I was wearing short trousers and there was no force behind it so it dribbled down my bare leg; I was amused at my discomfort and thought to myself, "That jolly well serves you right". Is this sort of reaction, even in principle, programmable? Are animals capable of irony?

Is it knowing of the latter sort, which of course colours all our knowing, which essentially distinguishes us from animals and which has been used by many philosophers to demonstrate that man has an immaterial element in him? The answer would seem to be, 'Yes and no'.

Bernard Lonnergan[8] in his huge work[9] started his enquiry into the nature of understanding with the dissection of the dawning of an insight, the moment of understanding which cannot be forced and which makes us smile, (or even leap out of our bath and shout with joy), the seeing of a joke three days after we have heard it and when we are thinking of something quite different. Cerebral activity is clearly going on all the time and some connection has been made and this enters our conscious awareness and is accompanied with a certain release from tension. This release, this feeling, sometimes of intense pleasure, no doubt has an evolutionary root and will be shared by us and animals and even, in a modified form, by plants and ultimately by all matter. It does not, it would seem, need any additional ingredient. The compulsion to search for understanding (heard in the incessant Why? Why? Why? of a child) and the satisfaction when the connection has been made, the answer found—often, incidentally, the wrong answer—can be satisfactorily explained on an evolutionary basis. Curiosity might occasionally kill a cat, but in forwarding its ability to learn and survive it saves a whole species. Insight is no more mysterious than having a headache. It is no less mysterious either. Knowledge has basically the same physiological base as sight.

We have some understanding of the physiological processes of sense perception but, rather like a Mandelbrot Set, every advance in understanding reveals further territories to be explored. An ever increasing understanding of chemistry is required and this in turn demands a deeper and deeper grasp of fundamental particle physics. It seems to me that if one is to try to come to an understanding of what it means to know, then some inkling of the means by which we receive information is a necessary pre-requisite. Philosophy and physics are inextricably linked.

Even when we have satisfied ourselves that we have some small glimmer into the way that our brain cells can be stimulated (or, and I often wonder whether this is the most that we can achieve, when we come to some vague insight into a possible solution[10]) there still remain two problems, one more or less in the same scale and one of an entirely different order. The first is the organisation, the chemistry and physics of the brain itself, of which in spite of the huge and rapidly growing body of knowledge, there is still precious little understanding. The second, of which as far as I am aware there is no understanding whatsoever, *not a glimmer of a glimmer of comprehension*, is the nexus between cerebral activity and an actual conscious perception, and between perceptions and understanding.

8. Whom my daughter refers to as The Learned Bonnergan.
9. Insight, Darton, Longman & Todd 1957.
10. A very similar thought is expressed by Paul Davies at the beginning of Chapter 2 in The Mind of God, Simon & Schuster UK 1992

And then there is the whole question of what the I is who perceives and knows. Assuredly it is not an atomic spirit located somewhere in the brain.

Whether science can bring any understanding of the underlying questions raised by metaphysics (including, of course, the validity of such questions in the first place) might be a different matter. By metaphysics I do not mean the integral heuristic structure of proportionate being[11] (at least, I do not think I do) but rather the science that underpins all sciences: the science of every science. For instance, an examination of Ockham's Razor—Entia non multiplicanda sunt praeter necessitatem or very loosely translated, 'The simplest explanation is the truest' is a matter for metaphysics rather than physics, biology, geology or geometry. It is claimed by evolutionists of the school of Richard Dawkins or Daniel Dennett that full and satisfying explanations of our understanding of logic can be found in Darwinism. It is forcefully argued by mathematicians such as Roger Penrose that the brain is a physical system (how it evolved is not really relevant) but that there are aspects of its actions that cannot be simulated computationally and that non-computational actions lie outside the currently known or assumed laws of physics.

My own view is that a wholly materialistic explanation is, to say the least, unconvincing. The reasons for this are discussed in Chapter 5.

All sense stimuli require the transference through space of some form of energy from the stimulator to the receptor: photons from the tree to the light cones in the retina, molecules from the rose bush to the sensitive cells in the nasal lining, electrons from the light socket to a variety of cells in the skin of forefinger and thumb and I know not what from the boxing glove to the chin or the stilton to the tongue but I am sure that in each case some physical transference across space, even if it is only of a tiny fraction of a millimetre, takes place. To achieve any understanding of this process needs therefore some inquiry into the nature of energy and space even if this only removes some prior assumptions without giving much insight.

I do of course, as has previously been said, recognize the complexity and shortcomings of language and understand that it must be used with what precision and refinement are possible and that we must always be aware that we can so easily be seduced into thinking that we are using a word with precision when in fact it has precious little content. Nevertheless I also firmly believe that the questions are real and that satisfying answers can at least be glimpsed.

I do not claim that I have found them, but in the meantime here is a seed of thought, a hint at a possible solution to the one of the questions which has so bedevilled philosophers through the ages:

The unperceived tree is incomplete in a manner somewhat analogous to the music stored on a compact disc.

---

11. Insight XV 1

CHAPTER 2

THE PERCEPTION OF SPACE, TIME AND MATTER

PART 1 SPACE AND TIME

Space and time are the things that most puzzle me in life, yet as I hardly ever think about them they might be said to puzzle me least.[1]

Although we are born with hugely complex central nervous systems, organised to grow and develop in human society so as to become capable when mature of the full range of human activity, walking, talking, thinking and loving, man needs a great variety of external sense inputs if he is to grow fully. A baby needs food and warmth and love; it needs to see and hear and taste and touch and smell, to hear language and music, see the sky and trees, smell flowers, feel fur and taste honey. All our sensual data are experienced in space and time and perhaps it will be well to start our enquiry into what is meant by "knowing" with a look at our current understanding of them.

Space, we think, is easier to understand than time because it seems obvious to us. It is there, and we inhabit it; It is what comes between me and you, between us and the moon. All right, but let us examine this a little more closely. It cannot be nothing strictly speaking because if nothing separated you from me (and I do so hope that nothing will separate us) or us from the moon, then we would be contiguous.[2] Is this just verbal cleverness do you think, or is there some substance behind such a simple idea? If there is some substance is this substance

1. Possibly Charles Lamb

2. Descartes expresses an identical thought in his Principles of Philosophy II 18: he beat me to it by some 350 years or it might be that I read it many years ago and it stuck in my subconscious.

space? What separates us from the moon? Nothing is clearly not a sufficient answer and if the answer is space then space cannot be nothing. Time is even more puzzling than space because it is constantly on the move like shifting sands. We cannot stop it to examine it; we can neither go back nor forward. Did time have a beginning or was St Augustine right when he said that there was no time before time began. (He also said that when asked what God did in the countless aeons before He created the world some one replied that He was fashioning a hell for people who asked such questions.)[3] Has time run evenly for all time? Is there any way of knowing? Is our perception, our experience of time a product of our nervous system—given a different one, could time be perceived as a colour? A changing colour? A cyclically changing colour but not in time? Is change possible without time?

Is the notion of matter any simpler? Let us take the matter with which I am most intimately connected—myself. When I rub my fingers together it is obvious that there is something real, something solid, there, but it is not as straightforward as that. When I look at my thumb, when I rub my thumb and forefinger together, nerve cells trigger minute electrical impulses to the brain and at a finite time after the rubbing or looking I feel or see a thumb and a forefinger. Given a different nervous system I would experience a completely different sort of thumb. We cannot imagine things which are wholly outside our experience, such as an entirely different colour, the colour of ultra violet for instance. That it could bear very little relation to my present experience can be understood by imagining (this might take some effort) that the nerve cells in the skin of our fingers responded only to heat and that the brain turned the received signals into a sensation of sound and further that the nerve cells in the eye operated on a radar system like those of a bat and that these signals were perceived as a smell. The thumb would be perceived as a greater or lesser noise depending on whether it was hotter or colder than the ambient temperature and with a fluctuating smell as it was brought nearer to or further from the eyes. What goes for things outside myself also goes for me: a refinement of solipsism and an understanding of what Descartes meant by his Cogito: that the only thing of which he was absolutely certain was his own consciousness.

The walnut tree in my garden quite clearly exists apart from me and I can investigate its constituent parts down to the atomic and sub-atomic level and describe their interactions in terms of mathematical formulae. It will be found that these relationships are stable throughout the universe and that they allow us to predict the behaviour of many things with astonishing accuracy, which is the use and the glory of science. The investigation cannot enquire into what an electron is in itself or in what the tree is in itself (if indeed these are meaningful notions) but it can find out a great deal about how the various constituents of a tree interact and the invariant patterns that they follow. Man can use this knowledge to the great benefit of mankind and of the living environment generally; it can also be used for the destruction of both.

3. Confessions XI.13

As a means of predicting the behaviour of a vast range of objects from the basic constituents of sub atomic particles such as quarks, to whole galaxies, relativity and quantum mechanics provide answers with astonishing accuracy. These theories will be examined in sufficient detail, it is hoped, to be made intelligible to the non specialist and a simple but sufficiently thorough presentation of the relevant maths will be included in an appendix to enable any one to understand the critical points at issue. This is necessary because they do have a huge bearing on our understanding of what the universe consists and any philosophy which leads to a logical contradiction or denies facts of observation is worthless. It must be emphasised that this is not in any way a text book. No attempt is made to cover much of the ground that is to be found in the very extensive literature, both popular and technical, which is readily available in all good bookshops and libraries. Only a barebones of the theories will presented but not an eclectic one: the aim is only to promote real understanding of the fundamental underlying principles. This is sufficient, I believe, to remove the misunderstanding of the nature of space and time that helped to lead to Kant's contortions but bring us no closer to an understanding of perception.

Space and time are clearly connected at least in so far that we cannot adequately describe a location without specifying a time. Only one car can occupy a particular place on a road at any given time. "Meet me at three o'clock," or, "meet me in Regent Street," are useless instructions, one without the other. And Regent Street lies on the surface of a revolving globe which itself is orbiting the sun which is sweeping through space (whatever that might be) and time ticks on remorselessly.

Would it be possible to define a location in terms of distance and direction (universal polar co-ordinates) from the location of the big bang if there was such a thing, and absolute time measured from its instance? Is there any meaning in this notion or that this point could itself have moved over the millennia? Has time remained uniform? Does distance remain uniform? Have these questions themselves any meaning or relevance?

If there was only one object in the universe, could it move? Once again the question is neither as simple nor as barmy as might first appear. First one would need to clarify what is meant by "one object". One space ship equipped with rocket motors could be accelerated, but once this happens there will be more than one object unless we treat the rocket and its expelled fuel all as the single object, and then in each object, however small one might choose it to be, there are always constituents. While cruising along in your space ship you are not aware of movement, you are not aware of gravity, but if the space ship were to be accelerated you would be aware of a force acting on you in the opposite direction to that of the acceleration. If this was at right angles to the floor of the space ship and at a rate of ten metres per second per second, you would not know whether you were stationary in a terrestrial gravitational field or accelerating through nearly empty space.

All our experience is of a three dimensional reality and a passing instant. Our concept of space is largely dependent on our mode of sensory perception. I wondered earlier whether given an entirely different sensory apparatus, our perception of space would be entirely different: entirely, utterly and completely different? Or is it possible that our mode of sensory perception is the only possible one: that the marvellously intricate interaction of the constituents of a tree, and of these constituents with my eyes are both part of an all embracing whole?

The time has come for at least a temporary end to these general speculations and to start to examine what we know or assume about the nature of space. Newton and his contemporaries believed in a space that was a finite or infinite three-dimensional homogeneous frame of reference with a uniform flux of time operating throughout the whole universe. Towards the end of the nineteenth century this view began to be questioned and ideas evolved and came to fruition early in the last, in the two theories of Relativity (Special and General) proposed by Einstein. These have profoundly modified—I am tempted to say mudified—our understanding of space and time. They are far from easy (and possibly impossible[4]) to understand fully; they operate on the very edges of our ability to understand yet no experiment yet devised has been able to show either theory to be false, and countless thousands of observations have been in full accordance, to a startling degree of accuracy with the theories—accurate to at least one part in ten million million.

The Special Theory states that the laws of physics—the patterns of behaviour of bodies—hold good in any inertial system: that is, one which is moving at a steady speed with respect to an observer. A good illustration of what is meant by saying that a system is inertial would be to imagine standing a penny on edge on a table. If it stays upright then you and the table are in an inertial system. The standard example is a smoothly running railway carriage and an embankment, in both of which situations the penny would remain standing. Within the carriage all the laws of physics hold good, as they do on the embankment. The Special Theory concerns and quantifies the modifications to measurements of space and time which have to be made when they are observed in a body which is moving at a uniform speed in relation to the observer. The concepts are difficult (and over a long period of time I have found them repeatedly difficult—perhaps I have never really understood them) but the mathematical analysis is comparatively simple. It involves the notion of the interdependence of time and space and leads to the equivalence of mass and energy.

A fundamental shortcoming of the Special Theory arises from the fact that changes take place in the carriage which was initially at rest in relation to the embankment which are not experienced by a person on the embankment. At some stage the train has to be accelerated and during this period the penny in the carriage will not remain standing whereas the one on the embankment will. Real forces are experienced in the carriage and there is an inescapable conclusion that it has really moved in relation to the embankment. This leads to the hopeless situation whereby the relativity of motion postulated by the Special Theory now

4. Theories are impossible to understand to the extent to which they are incomplete or mistaken.

has to contend with some form of absolute accelerations, which in turn require fixed reference points.

The General Theory overcomes this problem. It extends the equivalence of inertial systems to all systems by equating mechanical accelerations with gravitational fields. It overcomes the anomaly and in doing so provides a theory of the structure of space and time whereby they become a property of matter. No matter: no space, no time. To St. Augustine's, "There was no time before time began"[5] can now be added, "there was no space before space began". the maths involved in the derivation are much more difficult than the elementary maths required for the Special Theory. For a good understanding they are however regrettably necessary.

SPECIAL RELATIVITY

Space and Time

When it became clear in the nineteenth century that electromagnetic radiation had certain wavelike properties that were propagated through the vast regions of interstellar space a search was instituted to discover the nature of the medium which carried these waves: the ether. Entirely negative results of experiments carried out by Michelson and Morley designed to detect ether drift by measuring differences in the speed of light going with the prevailing current and against it led to the thought that the velocity of light relative to any observer was independent of the relative velocities of the emitter and the observer. Einstein's first really seminal intuition was that wherever a laboratory was set up on earth, or on any of the planets, or indeed anywhere in the universe, the laws of physics would remain the same and for this to be possible it was necessary for the velocity of light to be constant with respect to any observer. That this is so is made a little clearer if we consider talking to one another in a house, or in a car or in an aeroplane or in a space capsule on the way to Mars or sitting on Mars. In each case we can talk quite naturally. The laws that govern the propagation of sound and those that govern the reception are identical in each. In a similar manner one would expect a ray of light to be carried along inside the car or plane or capsule and that the results from any optical experiment would be unaffected by the location. For this to be so, if light was transmitted through a medium, the ether, then the medium itself would need to be carried along inside the capsule at a differential rate to the medium outside, in the same manner as the air inside a car or train is. This is quite contrary to the accepted notion of a vast unalterable space in which the universe exists. The idea of little bits of ether being carried around is as absurd as visualising space itself as being compartmentalised and carried around by the various compartments. If a ray of light inside the space capsule behaves just like one passing by on the outside, and the one inside is not

5. Confessions XI. 13

being carried along (like the sound wave is) then it follows that the velocity of light is unaffected by that of the emitter or receiver, all in conformity with Michelson-Morley.

The implications that follow from this by now extremely well tested observation of the velocity of light (that it is independent of the relative motions of the emitter and the receiver) are quite extraordinary.

A car travelling at sixty miles an hour down the motorway following another car travelling at sixty miles an hour does not move relative to it. If they are travelling in opposite directions they have a relative velocity of one hundred and twenty miles per hour, and should they touch each other the consequences are likely to be catastrophic. If you fire a low velocity cannon ball at the back of a passing express it might not even reach it, but if fired into an approaching express it will have a devastating effect.

If you shine a torch at the car you are following, at the car you are approaching or at either of the express trains or at a high velocity bullet speeding away from you or indeed at anything whatsoever, the light, the photons, will reach or recede from each at exactly the same velocity, three hundred million metres per second. By following the logical implications that flow from this extraordinary fact and in a fit of intuitive genius linking the result to earlier equations worked out by Maxwell and Lorentz, with his own notions of the symmetry of the universe, Einstein was able to forecast the interchangeability of matter and energy: that matter can be turned into energy or energy congealed into matter, in his simple equation $E = mc^2$ where E is energy, m is mass and c the speed of light.

(The amount of energy which can be unleashed from a body of mass m if it is turned totally into energy is its mass multiplied by the square of the velocity of light. If m is measured in kilograms and c in metres per second then the energy will be in joules. One joule is the energy needed to accelerate a one kilogram lump (a large packet of butter) to a velocity of one metre per second over a distance of one metre on frictionless ice. One kilogram of matter converted entirely into energy would release nearly ten billion billion or ten million million million joules—enough to drive a million cars flat out five hundred times right round the globe—all this from a litre of water or a small bag of potatoes or a large packet of butter.)

It was felt at the time that he was dealing in mere speculation in an abstract academic argument but nevertheless an argument very difficult to refute, so it attracted a lot of serious attention. At the time indeed to some extent he was, for there is no straightforward analytical chain of logic and there were no possible experiments to test the theory but Hiroshima and Chernobyl on the one hand and (so far) the hundreds of satisfactory nuclear power stations around the world on the other have shown that it was far from mere academia. Matter is, we think, easy to conceptualise but both mass and energy are very difficult, and both are aspects of the same stuff: Aquinas's prima materia.[6]

It cannot be too much emphasised that the basic assumptions, or, if you will, laws, on which the whole of physics rests have all been tested hundreds and thousands of times and have never been shown to be false. Such modifications they have undergone in the light of modern physics, such as relativity and quantum theory have served to deepen our understanding of them.

The chain of reasoning and the intuitive assumptions that underlie the Special Theory of Relativity are as follows. There are two what might be termed philosophic assumptions which underpin all science and five scientific ones. The philosophic assumptions are the belief that the universe is susceptible to rational questioning and in Ockham's razor: that the simplest explanation is the truest. The scientific assumptions are:

1. Conservation of Matter
2. Conservation of Energy
3. Conservation of Momentum,
4. Newton's First Law
5. Newton's Corollary.

The conservation laws state that matter and energy can neither be created nor destroyed and that in any collision total momentum (the product of the mass and velocities of each of the colliding objects) will remain constant. Newton's First Law states that any body will remain at rest or in uniform straight line motion unless acted upon by an external force. Normally forces such as gravity and friction, or air resistance acting on everything make direct experiments difficult but in, for instance, a freely falling space capsule the conditions exist where bodies will nearly act in conformity with no "unless" with respect to the capsule—nearly because some atmosphere will be present and there will be gravitational forces from the capsule itself and from other objects within it. The capsule will act as a frame of reference within which the law could be tested. Any frame of reference where Newton's first law accurately describes the behaviour of any body within it is known as an inertial frame.

Newton's Corollary is that in any inertial frame all the laws of physics are identical. For instance, large cruise liners have tennis courts and billiard tables and neither you nor the tennis balls nor the billiard balls are aware that the boat is moving relative to the land, but if viewed from the land the motion of men and balls would be quite complex. The fact that boats travel relatively slowly is immaterial: I heard of a Boeing fitted with a squash court. Of course while the boat is speeding up or slowing down, or caught in a severe storm, or while the aeroplane is taking off, no games will be possible because the frames in which they are taking place, the boat or the plane, are accelerating or decelerating: they are not inertial. A corollary of this axiom (and one that has been well tested) is

---

6. Matter which has no form whatsoever. Summa Theologica 1 LXV1 1 cf Page 32 under.

that the velocity of light in a vacuum is a constant—approximately 299.792 million metres per second—in all inertial frames.

From these two assumptions and five laws the special theory of relativity was worked out with all its counter intuitive prognostications: the equivalence and interchangeability of matter and energy, of light being affected by gravity, of time and distance being dependent on the relative positions of observer and observed, of the nature of space and the space-time continuum. Much preliminary work had been done by Maxwell and by Lorentz and the experimental work with respect to the velocity of light by Michelson and Morley, and a whole host of others but the derivation of the theory is due to Einstein. It has been reworked over the last century, starting often at first principles. Rigorous modern analytical approaches such as that of Wolfgang Rindler in *An Introduction to Special Relativity (OUP 1997)* can be consulted but this, though admirably clear, assumes a knowledge both of maths and physics of a second or third year physics undergraduate. An attempt to give as nearly as rigorous but very selective barebones of the theory is presented in the second appendix, using only school mathematics. Because most of us have forgotten any maths we were ever taught this too is recapitulated from first principles and assumes no prior knowledge. The hope is that a real understanding of special relativity will be promoted and if possible to enhance the understanding that there are great depths to be understood: that things are not as simple as they seem, at least if we are looking for simplicity in the make up of the external world. I think a more accurate articulation of my real feelings is not that the world is less simple than it seems but rather that simplicity is not simplistic: great simplicity has great depth. A good analogy is the advance made by Copernicus over Ptolemy—a much simpler system but one which requires or leads to a much greater depth of understanding.

The fact that light rays travel at a constant velocity in all inertial frames is impossible to visualise. The rays from a torch shining down the corridor of a high speed train and those from one on the platform of a station through which it thunders will keep pace with one another. Yet if we measure their velocity relative to the moving torch and the stationary one they will be found to be identical. The only explanation of this is that were we to measure the length of the train from the platform the train will be shorter than when measured inside the train. when measured from the platform. If we knew its speed we could time its passage and compute its length and compare this with the length measured within the train, either by running a tape measure down it, or by timing the return trip of light from one end to a reflector at the other and back. The light would have to travel further along the track than inside the train unless the train shortens.

Similarly if we time the rays of light, the clock in the train will need to run more slowly than the one on the platform.

The actual changes are equal to $\sqrt{(1-v^2/c^2)}$ where v and c are respectively the relative velocities of the objects and that of light.

At ordinary terrestrial velocities the effects are negligible but with the advent of much higher velocities in space travel together with very accurate atomic clocks they are just measurable and confirm the theory. Even at six million miles per hour the difference in time would be less than half a minute a year.

In a moving object, relative to a stationary one, everything is slowed down right to the motion of electrons round atoms. There is no such thing as absolute time. Time on the earth, travelling at huge velocities away from a distant star would be much slower as measured from the star compared to time on the star itself. A clock sent into space at a very high velocity will show a smaller lapse of time on its return than its earthbound partner. A twin sent into space at sufficiently high velocity would only age ten years while the stay-at-home aged twenty. There is an asymmetry here: the twin would appear to grow shorter while in space but would grow taller again as he returned to earth; his rate of ageing would approach that of his twin but this would not be retrospective.

Mass

The second of Einstein's seminal intuitions arose from the thought that if a ball is thrown horizontally across a room it will describe a parabola: its horizontal velocity will remain constant but gravity will cause it to accelerate downwards. If a ball is thrown in a space-ship coasting freely in space it will travel straight across it; if the ship is accelerating at ten metres per second per second it will describe a parabola identical to the one in the room on earth. If a ray of light is shone externally across a window in a spaceship which is accelerating upwards it will also describe a parabola as far as some one inside the ship is concerned. If some one switches on a torch inside an accelerating spaceship would it also describe a parabola? If some one switches on a torch in a room on earth, will its light trace a parabola? Of course it would, thought Einstein. Gravity and acceleration are identical and light should be affected by gravity. A person inside the ship does not know whether he is accelerating in open space or is stationary in a gravitational field. Nor, thought Einstein, would the ray of light and this should therefore behave identically whether passing across an accelerating object or being subjected to gravity. Light should be subject to gravity in exactly the same way as other objects, and this is subject to experimental proof, with light or other forms of radiation passing close to the sun on their way to us on earth and so it has proved with increasing precision over the years. The equivalence of as it were mechanical accelerations and gravity is the subject of the General Theory of Relativity and we are getting ahead of ourselves.

From the analysis of the effect of space and time measured from one object to another moving at a constant velocity came the intuition of the equivalence of mass and energy. There is no clear logical derivation of this but there are plenty of pointers before one has to invoke the Law of Elegance.

Energy is a measure of work: it is what is needed to push a car along or speed it up. If the car is on a slight down-slope such that it needs only a finger touch to overcome the friction in its wheel bearings and move it but it would stop once the finger is removed, no energy is needed to move it but energy would be needed to speed up the car. (In numerical terms, $e = {}^1/_2mv^2$ where e is the kinetic energy of the body, m its mass and v its velocity.)

There is no clear logical route from the derivations of the relativistic formulae of the transformations of time and distance (with relativity factors of $\sqrt{(1—v^2/c^2)}$ to the famous equation of the equivalence of mass and energy, $E = mc^2$ but the pointers are as follows:

Firstly an assumption of a relativistic formula for the transformation of mass;

Secondly a look at the effect of this on the kinetic energy of a body

Thirdly by an imaginative leap the equivalence of mass and energy.

Mass is a measure of inertia[7], the measure of the resistance of a body to be accelerated when acted on by a force. It is a measure of the density of a body, and is not to be confused with weight which is a measure of the effect a gravitational field has on a body. A cubic metre of water would have the same mass on the moon as it does on earth although it would have only a fraction of the weight.

The relativistic effect on mass has the familiar form: $\sqrt{(1—v^2/c^2)}$ but in this case whereas the length of the moving object shortens, its mass increases: as its volume diminishes its density increases. Here we have a strange fact. The slowing of time is reasonably straightforward to derive given the extraordinary fact that the velocity of light is the same in all inertial frames. It is difficult to verify because the practical effects on time are tiny at even comparatively huge terrestrial velocities (say, one hundred thousand miles per hour) but experiments with very accurate clocks have been done and the theory borne out. The shortening of length is easily derived from the first but unlike the clocks has not yet, I think, been able to be verified by experiment. The relativistic transformation of mass requires all sorts of shenanigans to derive; the famous equation $E = mc^2$ is quite easily derived from the relativistic equation, as demonstrated in the Appendix and the consequences are very large. They have been spectacularly verified by countless thousands of occurrences in atomic power stations, by atomic bombs and by innumerable laboratory experiments.

In practice only a tiny proportion of any substance is converted to energy. This is not so important, as the energy liberated by relatively small changes of mass is enormous. A diminution of mass of one gram will provide enough energy to lift a weight of a hundred tons[8] (two heavily laden articulated trucks) from sea level to the top of Mount Everest.[9]

---

7. The New Shorter Oxford Dictionary defines mass in this context as, "The quantity of matter which a body contains, as measured by its acceleration under a given force or by the force exerted on it by a gravitational field" and inertia as, "The property of a body, proportional to its mass, by virtue of which it continues in a state of rest or uniform straight motion in the absence of an external force".

8. A metric ton (the tonne or euroton) = 1000kg = 1.016 UK or long ton For most practical purposes they are the same.

9. In 1920 when Rutherford first managed to release some energy by nuclear fission of a nitrogen atom, the popular example was that a gram of matter was equivalent to three thousand tons of coal.

E (joules) = 1/1000 kg x 3,000,000 m/s x 3,000,000 m/s = 9,000,000,000 joules. Mount Everest is approximately 9km high and the energy needed to lift a weight of one hundred tons to this height = mass x acceleration x distance, where the acceleration is that of gravity or approximately 10 m/s/s giving an energy requirement of (100x1000)kg x 10m/s^2 x 9000m or 9,000,000,000 joules.

The Space-Time Continuum

Finally, we need to look at our notions of space and the relationship of space and time.

While driving you can chat to your passenger or listen to the car radio much as you do when the car is stationary. Sound consists of fluctuations of pressure in the atmosphere travelling at around seven hundred miles per hour from the emitter to the receiver. The perception of a note of Middle C is the end result of a minute pressure rise and fall two hundred and fifty-six times per second transmitted through the air and acting on our ear drums; its loudness depends on the size of the fluctuations; its tone depends on the smoothness of each fluctuation. The reason that we can talk in a moving car is that the air inside the car is carried along with it, and, relative to the car, is at rest. Open the windows wide and the same is no longer true. If our passenger has a cup of coffee it is a matter of common (and sometimes painful) observation that while the road is fairly straight and our speed fairly constant the coffee will remain in the cup. The same is not true if we brake suddenly or go round a sharp corner.

From hundreds and thousands of observations of this sort a general truth has come to be recognised: that objects in any closed non-accelerating system behave in identical ways when observed from within the system.[10]

Take as your closed system the planet earth: we move round the sun on a ninety million mile circle once every year. The circumference is five hundred and sixty million miles, giving an orbital velocity of seventeen miles per second relative to the sun, and a difference of thirty four miles per second to a star in a direct line through us from the sun in say June and December. Light too consists of some sort of fluctuations travelling from the source to the eye, at a much greater speed than sound (one hundred and eighty six thousand miles per second) and very much greater frequency (almost ten thousand million million oscillations per second). It travels vast distances through space from distant stars to us and the question of the nature of space itself must be raised. When a car moves the air inside moves with it, leaving a temporary vacuum behind. The notion that a car carries its own space through space leaving a temporary spaceless region behind it offends our intelligence: we think of space as a more or less unlimited three-dimensional nothingness. Through this nothingness material

10. cf Newton's Corollary

objects (planets and rockets) can travel and have varying distances from other objects.

Let us probe this idea of nothingness for a moment. Some years ago I had the thought that if literally nothing separated us from the moon, then we should be touching it, and was pleased with the originality of the thought, only to find later that Descartes had pre-empted me by a few hundred years. I am sitting here (not in a car but an armchair) and a radio is playing in the background. I understand, more or less, that I am surrounded by more or less still air (the house is draughty) and that fluctuations in its pressure convey the sound to my ears. I can only too clearly feel the draughts. I am also aware that the radio is activated by radio waves, which are all around me, together with a whole host of other phenomena: heat waves, (not heat-waves, it is January in England) light waves, gravity waves, magnetic waves and a lack, inside my head, of brainwaves. All this activity is taking place in empty space, not only here but across the universe. The emptiness is by no means empty.

We have receptors which have evolved (by chance) to be sensitive to a tiny and highly selective proportion of all that goes to make up the universe. We are unaware of light just beyond the violet end of the spectrum until we are sunburnt by it or develop cancer; we are totally unaware of radio waves, x-rays, gamma rays or gravity waves. Had other receptors evolved space would not seem empty but solid throughout the universe: the things that go to make up matter (atoms and electrons and the like) are no more solid than the range of electromagnetic phenomena that pervade space.

Space is the aggregate of these phenomena.

Outside the universe there is nothing. The universe does not exist in space: space is a property of matter.

When space is said to be warped what is meant is that the arrangement of all these various fields is altered: for example, the path of light between two objects is changed. A straight line is the line taken by a ray of light: the line through the sights on a rifle from the object to our eye. If the ray passes an object it will be pulled towards the object by its gravitational force but it will still look perfectly straight as far as we are concerned.

For the ray of light to be able to move at all, time is required: as has already been pointed out, time and space are part of a single whole. For the universe to be able to expand or contract, for space itself, that is, to grow larger or smaller, time is a necessary constituent, and all three measures of the basic constituents of the material universe, distance, time and mass are subject to the relativistic formula $\sqrt{(1—v^2/c^2)}$ energy is tied in by the famous equation $e = mc^2$ and only gravity escapes the net for the moment. When a satisfactory unified theory does finally emerge then as Relativity and Quantum Mechanics profoundly modified and added to Newton's grand synthesis, so it seems probable that further equally profound modifications will be brought to light. Even when a unified theory is formulated it will not be the end of the matter. Just as there are no limits to what we can aspire to understand, there are, I believe, no limits to what can be understood.

PART II MATTER

Take a simple, sharply defined solid object like a ball bearing and look at it closely. If we hold it in our hand it appears as a small spherical shiny object. If evolution had taken a different turn and our eyes and nervous system were differently constructed so that they acted as powerful microscopes, the surface would look much more like a relief map of the Alps. If our sensors acted like an electron microscope huge voids would appear: the apparent solidity would be revealed to be largely empty space, and the static nature of the ball would be seen to be a pulsating maelstrom. If we were able to sense the various forms of electromagnetic radiation the voids would fill up and the surface of the sphere expand far beyond its visual presence.

We have the eyes that we have and our visual perception of the world comes from what we can see, but we can look at things through magnifying glasses (or even, if our eyes are not perfect, through spectacles) and see things more clearly; although we cannot look through an electron microscope it does provide us with an image of an object far smaller than could be seen through the most powerful optical microscope. We can guess at theories of patterns of behaviour and test and verify these by experiments. All this lays down memories (actual physical modifications of cells in our brains) and these to some extent alter our primary visual perceptions.

Knowledge of the universe can only come from observation of how things interact, and the very act of observing is itself an interaction.

As far as we know the whole of the material universe is composed of matter which can appear in one of two fundamental forms, which under some circumstances are interchangeable. On the one hand there are bodies which have mass and can have relative motion. All these objects are composed of atoms or their fundamental constituents such as electrons, protons and neutrons. On the other hand there is radiation like light which always has an invariant velocity, relative to which all other matter is at rest. Neither the bodies which have mass nor radiation can be accurately thought of as particles and neither can be described only as waves. Both exhibit some characteristics of what we understand by a particle and both exhibit some wave like properties. The *prima materia* of Thomas Aquinas is fascinating in this respect. As I understand it, he envisaged prime matter as, as it were, the single completely unformed totality of the material universe, in his paraphrase of Aristotle: nec quid nec quantum nec quale nec aliud quidquam eorum quibus determinatur ens[11] but capable of receiving this or that form—a horse, a tree—in Copleston's phrase, the indeterminate substrate of

11. Being, neither what nor how much nor of what sort nor anything else whatsoever by which it might be described: Metaphysics, 7, 2

substantial change.[12] Prime matter is always an abstraction; it cannot exist without some form but in our current state of knowledge (belief?) the closest we can get to it is perhaps in the first micro-second after the big bang. Since then the laws of the conservation of mass and energy have applied as far as we know: matter and energy in their totality have neither been created or destroyed. It has crystallised into unimaginably large numbers of particle-waves or wave-particles which combine and disassociate and change from one to the other ceaselessly. They, the protons, neutrons, electrons, photons and all the other elementary particles are perhaps (with one major exception) the only true substances of Aquinas's world. Atoms are of the order of one ten millionth of a millimetre in diameter; protons and neutrons which form their core have masses of 10^{-30}kg[13]. That is to say, in a litre of water, with one proton to each hydrogen atom and eight protons and eight neutrons to each oxygen atom, giving eighteen protons and neutrons to each H_2O molecule, there are something like 10^{30} (a million million million million million) molecules. Electrons, which form the outer "shells" of atoms are ten thousand times lighter than protons or neutrons.

Electromagnetic radiation ranges from γ-rays with frequencies of 10^{27} cycles per second through light at 10^{14}, FM radio at 10^{8} to electrical disturbances at one hundred cycles per second, which is in the audible range. Middle C is 256 cycles per second or, which is the same thing, 2.56×10^{2}.

> *All forms of action and interaction require a transfer of energy and as far as we know this is only achieved by the emission or absorption of radiation: of discrete particles or waves of energy.*

The particles or waves of energy are called photons and apart from nuclear reactions they are emitted or absorbed only by electrons. Electrons are the smallest particles or waves of matter having mass. All normal physical, chemical and biological interactions are the result of absorption or emission of photons by electrons. When I sit and join my hands and tap my fingers on each other, the pressure I feel is the result of a huge number of photons being absorbed by electrons in the cells of my skin, which in turn emit photons which are absorbed by further electrons in my central nervous system and finally by emission and absorption of photons and electrons in my brain.

The smallest object we can see with our own eyes, through an optical microscope, is of the order of one hundred thousandth of a millimetre; an atom is a thousand times smaller and an electron millions of times smaller. Because we cannot perceive atoms or electrons or photons we have no imagination of their nature and no language to adequately describe them: no meaning can be given to them in terms of sight, sound, touch, taste or smell. The following overview of quantum physics is therefore only an attempt to squeeze into words that which cannot be squeezed, as an exercise in heurism. A photon is a particle travelling through space always at the invariant speed of light; there is an associated

---

12. Copleston History of Philosophy 2, 326. Burns Oates 1960

13. 10^{30} means ten multiplied by itself thirty times, 10^{-30} is a fraction of one over ten multiplied by itself thirty times.

spherical spreading out from its point of emission, somewhere on which the photon is located. The photon should not be thought to be spread out over the front: if the front is intercepted by an electron there is a chance that the photon will be absorbed by it but until this happens no precise meaning can be given to its location. The photon is pulsating; the frequency of the fluctuations determine the energy content of the photon and how it appears to us—as a radio wave or light ray, x-ray or γ-ray. When it is absorbed it ceases to be a photon so the only time that the location of a photon can be measured is at the instance that it ceases to be, when its wavefront collapses to a point. The chance (or probability) of the photon being intercepted is in accordance with certain probability laws which are the essential subject of quantum physics.

These probabilities are expressed in terms of complex numbers, that is, numbers which are composed of two parts one of which involves the use of the square root of minus one. The absorption (or emission) of a photon by an electron is accompanied by changes in the mass and momentum of the electron.

Particles which have relative motion have associated spherical fronts spreading out at the speed that the particles are moving. Unlike the case with a photon where we have no knowledge of its location within its front, the precise location of any particle in its wave front can be known to a certain extent but this is limited by the accuracy with which we know its velocity: the more certain we are of its velocity, the less we can know of its position and vice versa. Subject always to this uncertainty, the location of any particle in its wave front is a matter of probability, expressed as a complex number, which is the subject of quantum physics.

The same fundamental laws cover both forms of particle and hence everything in the universe.

The only form of interaction in the physical universe with the exception of gravity and atomic nuclear actions is the absorption or emission of a photon by an electron. This includes everything from electromagnetic phenomena to all physical, chemical and biological events; they are all covered by the same fundamental laws which have been increasingly tested for over half a century to astonishing degrees of accuracy and never falsified.

In the closing years of the nineteenth century the failed search for the medium that carried electromagnetic radiation led to the formulation of Special and General Relativity. These were tours de force of purely abstract reasoning which had to wait several years before a suitable event occurred, such as a complete 3

eclipse of the sun, before the theories could be put to a practical test and for decades before the equivalence of mass and energy could be demonstrated. At the same time, at the turn of the century, it became apparent that Newton's laws broke down completely when applied to the sub-atomic world and theoretical development was forced on physicists.

It was known that energy could be transferred across space by means of an electromagnetic wave, with the type of radiation being dependent on the frequency of the wave. Naturally occurring frequencies vary from electrical disturbances at a hundred cycles per second to radio waves at between one million and one hundred million to light at between five hundred million million and a thousand million million depending on the colour and then X rays and γ rays at hugely greater frequencies. Equations for these waves had been put forward by Maxwell in the mid nineteenth century so that the electromagnetic potential (made up of the strength of a charge and its phase) could be calculated at any location and moment.

Three propositions in particular were made which led to the reconciliation of Newtonian theory with observation: in 1900 Planck put forward the idea that energy carried by radiation was emitted not as a continuum but in discrete packets, quanta; in 1905 Einstein proposed that light rays behaved not only as a pure wave form of radiation but also as a stream of particles and in 1924 de Broglie suggested that the behaviour of elementary particles, such as electrons, had certain wave-like properties. Two years later Schrödinger introduced his wave equation which became the fundamental equation in quantum physics. During this period and continuing into the thirties, a host of physicists developed the ideas, backing their results all the time with experiments and there arose a electromagnetic wave fundamentally different form of physics to that proposed by Newton which modifies rather than contradicts his laws much as in the case of relativity and which reduces to his laws in the world of ordinary experience. Whereas the theories of relativity were largely the work of one man, dozens made significant contributions to quantum theory. As with relativity, and to an even greater extent, the theory has been supported by hundreds of thousands of events and experiments to an astonishing degree of accuracy and has never failed. In the atomic scale world, in the development of lasers and of the silicon chip for example, it has been an indispensable aid.

Quantum Mechanics

Essentially the theory has a number of separate strands:

i) Energy can only be emitted or absorbed in discrete packets or quanta.

ii) A quantum of energy is proportional to its frequency.

iii) The behaviour of all matter, microscopic and macroscopic, including all forms of radiation, is in accordance with certain probability waves.

iv) As a corollary of this, all forms of matter exhibit characteristics both of particles and waves.

v) Because of this duality measurements which rely on one or other aspect cannot be made simultaneously to arbitrary degrees of accuracy.

From these strands, each of which will now be briefly discussed the formal structure of quantum mechanics evolved.

i) Energy can only be emitted or absorbed in discrete packets or quanta. Energy is always transmitted in discrete quanta although because the rate of emission is very fast in comparison with the speed of our perceptions it will appear as a continuum much as discrete knocks merge into a single sound as their frequency increases.

ii) A quantum of energy is proportional to its frequency.

The smallest packet in which energy can either be emitted or absorbed by a body, is dependent on the frequency of the energy wave such that $E = h\nu$ where h is Planck's constant and ν is the frequency of the wave. h has a value of $6.63 * 10^{-34}$ joule seconds.[14]

iii) The behaviour of all matter, microscopic and macroscopic including all forms of radiation, is in accordance with certain probability waves.

There are waves in the sea, where the crests can be stationary or moving; there are sound waves and radio waves, there are hair waves, heat waves, crime waves and brain waves (I could do with one of these). The essence of a wave, whether concrete or abstract, is that it is undulating. It has peaks and troughs. It does not generally have a definite location at any given time but rather it is present in a given space. It has a number of velocities associated with it: its front might be spreading in various directions at various speeds; its peaks and troughs might be replacing each other at varying frequencies at different locations at any given moment. It might be static and material as in undulating ground or stationary but made up of particles moving in a wave as in a standing wave in a water trough or in a violin string or even the movement of an arm when saying goodbye or the wave itself might be moving as in the ripples that spread across a pond when a stone is thrown into it or the pressure vibrations of air as a sound wave moves through it; it might be abstract as in a wave of reformatory zeal or a new wave of high technology that sweeps a nation.

If an oscillating plunger is placed in the centre of a pond it will cause ripples to spread out. The height of the water above or below the undisturbed plane at any location in the pond will vary with the distance from the plunger and the time: it is a function of both space and time. If the properties of the plunger and the water are known it is possible to form an equation which can be solved to give the wave height at any location at any time. It is the essence of a wave that as its height increases above a middle position the vertical velocity of a particle on it lessens until it comes to zero at the top of the wave after which it moves downwards. From the mid point to the top the particle has an increasing acceleration in the opposite direction to which it is travelling; it, as it were, brakes ever more sharply until it ceases to gain height. The rate at which it loses height then increases until it reaches its mean position but with a decreasing acceleration which becomes zero at the mid point, and so on. The velocity, the rate of

---

14. Just a reminder: one joule is the energy needed to accelerate one kilogram at one metre per second per second for a distance of one metre. It has the dimensions ML^2T^{-2}. One millionth is 10^{-6}; one millionth of one millionth is 10^{-12}. . . . 10^{-34} joules would be sufficient energy to give an electron an acceleration of around a millimetre per second per second. A light wave with a frequency of 7×10^{14} cycles per second will carry 5×10^{-19} joules per photon. A laser emitting light of that frequency in a stream of single photons would carry energy at the rate of approximately one joule per hour. One watt is 3600 joules per hour. A 40 watt bulb emits ten million million million photons per second: we are dealing with very large numbers of incredibly small objects.]

increase in height, is the first differential of height with respect to distance and time and is given the general symbol ∇ or more fully $\nabla(x,t)$ to show that the differential is both with respect to distance and time. The acceleration, the rate of increase of velocity, is given the symbol ∇^2 meaning the differential of a differential rather than a differential squared. The simplest form of a wave equation concerning the height of a wave with respect to time and distance would be $h(x,t) = - \nabla^2 h(x,t)$ which, being translated, means that the height, a function both of horizontal distance and of time varies in accordance with the negative acceleration of the height. When a point on the surface of a wave is moving upwards, it is accelerating downwards; when the wave is at its top the point is vertically stationary but has the greatest acceleration downwards; when it is at midpoint, it is moving fastest vertically downwards but has ceased to accelerate; when it is at the bottom it is again stationary but has maximum vertical acceleration upwards. The solution of this equation at any fixed location or at any fixed time is a pure sine wave.

A more general equation, taking into account the nature of the water (its viscosity and temperature, for instance) would take the form

$A \nabla^2 h(x,t) + B\nabla h(x,t) + Ch(x,t) + D = 0$ where A, B, C and D are specific factors.

In an electromagnetic wave there appear to be three quite distinct wave elements. First and foremost there is the photon itself, a pulsating packet of energy travelling through space at the speed of light, an energy field spreading out in all directions from its point of emission. Secondly there are streams of photons being emitted at a certain frequency with the photons in each stream having identical "internal" frequencies and thirdly there are abstract probability waves which allow us to forecast the behaviour of a photon (or stream of photons) in terms of location, momentum etc. Quantum physics is concerned with this third aspect.

The fact that we can only forecast the behaviour of photons in terms of probabilities is due to their discrete nature, and this will now be discussed. The reason why these probabilities are expressed in terms of waves is perhaps due to the pulsating nature of the photon itself: it is generally regarded as a fact of nature and it will not be discussed. (Beyond the vague notion expressed, I have no thoughts or understanding of the matter or indeed of matter itself.)

Every action requires some energy transfer and therefore every action is made up of a number of discrete quanta. Because of this all physical laws must be statistical and only probabilities can be forecast.

Let us examine a very simple analogy. Take delivery of a new roulette wheel, spin it and watch the ball settle on thirteen. Nothing surprising: just a matter of chance. Spin it again, and again number thirteen occurs. Two unlucky omens: well, the chance of this happening are about one thousand to one against and some thousands of new roulette wheels are delivered each year, so still nothing all that surprising. When the third and fourth thirteen turn up on the following throws the balance of the wheel would undoubtedly be questioned: it might be a matter of chance but the balance of probability must now be swinging in favour of a mechanical fault. We surmise—we know (how do we *know*?)—that although in any particular spin of a properly balanced wheel, the number on which the ball will come to rest is a matter of blind chance, over very many throws each number will tend to be equally represented: there is no blind chance in operation here. We *know* because all our experience and the accumulated experience of mankind has led us to believe that the universe is reasonable and there is no reason why one number should be favoured over any other.[15] Over a large number of throws all numbers will even out. We could in this instance reason that the degree of control over our hand movements is such that even taking as great care as we possibly can, the variation is such that we cannot control the interaction between the spin of the wheel and the throw of the ball to any number or group of numbers. If any particular number or group of numbers were favoured we would need to look for a reason. If the croupier was suspected of having refined his control, other croupiers could be used to test the evenness of the wheel. If the events still favoured the same numbers the wheel itself would be examined.

Any single quantum event (the emission of a single photon for example) is random in so far as we are quite unable to predict its timing, but a large number will form a pattern that can be predicted. So, in all physical actions, each made up of a few quanta in the case of individual sub-atomic events to huge numbers in the ordinary observable world, randomness is governed by the rationality of the universe. Newton discovered some basic patterns or laws governing the behaviour of bodies. Quantum physics modifies these laws to take into account their essentially statistical nature.

This is generally looked on as a fundamental shift from a causally determined universe to an indeterminate one: the collapse of determinism. It seems to me that this is a purely abstract argument and that nothing has changed in science itself: the practical application of physical laws has always been statistical.[16]

It might be argued that in the case of the roulette wheel it would in principle be possible to predict the outcome of any given spin provided that the initial conditions were known—the force of the spin, the direction and velocity of the ball and so on and so on. A closer examination must lead to the conclusion that to know the outcome with certainty the disposition of everything in the universe would need to be known: a slight earth tremor or an onlooker coughing could just tip the balance between one number and the next. Any one who has conducted any laboratory experiment will know that perfect conformity with any law is not possible: the law is always an abstraction from a smoothing out of

15. Unless it be 42, but that is not available on a roulette wheel.

16. cf Richard Feynman Lectures on Physics 1.39 "Is it not therefore fair to say that from the apparent freedom and indeterminacy of the human mind, we should have realized that classical "deterministic" physics could not ever hope to understand it, and to welcome quantum mechanics as a release from a "completely mechanistic" universe. For already in classical mechanics there was indeterminability from a practical point of view."

experimental evidence. I do not think that even in principle classical physics can ever provide actual predictions which are not expressed in terms of probabilities. In practice the results of experiments are inescapably conditioned by limitations of accuracy.

In the case of the emission of a single quantum it cannot even be argued, as far as we know, that it would in principle be possible to predict the outcome provided the initial conditions were known in that we do not know of the existence of any such conditions. We have come to the end of the chain at least as far as present knowledge is concerned.

Whether absolute chance is possible—that is to say, an event with no cause at all—well, I have my doubts.

The actual probability is always found to be the solution of an equation of a wave where the form of the wave is an assumed basic fundamental fact of nature. It cannot be derived or explained: it is just the way things are. Unfortunately it requires a mathematical notation which is generally beyond school maths and hence one that will be unfamiliar to the great majority of even highly educated people. It is known as Schrödinger's Equation; this is given together with an explanation of each of its terms, in Appendix 2.

iv) As a corollary of this, all forms of matter exhibit characteristics both of particles and waves.

From the fact that any physical interaction, from the movement of an electron from one orbit round an atom to another, to the earth circling the sun, and from the emission of a single photon to the radiance of the sun, is made up of discrete quanta which behave in accordance with certain statistical probabilities which are the solutions of wave equations means that all matter can—and indeed must—be treated as exhibiting characteristics both of particles and waves. It is only through its behaviour that we know anything about any particular form of matter. Movement in itself does not signify any action since it is relative.

The idea of a particle is fairly simple: it is a little solid something like a small ball bearing; its boundaries are tightly circumscribed and at any given time it has a specific location and velocity and its trajectory can be tracked. Whether anything in nature corresponds even remotely to this idea is another matter, and one that will be explored below. (You might object that it is quite clear that a small ball bearing is an object in nature that corresponds perfectly but this has already been exploded.)

When dealing with very small particles, like electrons or atoms or any particle which is too small to be observed directly even through an optical microscope, indirect means have to be devised if their movements are to be tracked. For instance, a stream of electrons will cause a patch of luminescence on the screen of cathode ray tube and by passing the stream through a magnetic field and observing the movement of the spot on the screen a knowledge of the path of the stream is gained. Similarly, by using a suitable permanent recorder, such as specially prepared photographic plates, the paths of single electrons can be followed at least in so far as their starting and final positions are known. Two troubles now arise. Firstly when we try to pin down intermediate positions, whatever form of experiment has been devised only probabilities can be assigned to the particle being in any particular location at any given time although for some locations and some times the probabilities might be one or zero: certainty or impossibility. Secondly if intermediate obstacles are erected which would cause waves to exhibit interference patterns in their final positions, these patterns are observed. If in the above experiments a stream of light replaces that of electrons exactly the same phenomena follow. If electrons were to be replaced by atoms the same patterns would follow but now the interference fringes would be much closer and more difficult to observe. If atoms were replaced by ball bearings the same would still be true but the interference fringes would be too close to be observable.

There is no essential difference in the behaviour of photons, electrons, atoms, molecules, bullets or planets.

All the laws of physics need to be expressed in terms of probabilities although once bodies of atomic or greater size are being treated the probabilities forecast only small deviations from Newtonian mechanics and by the time the objects are ones of ordinary experience the deviations are utterly insignificant and quite unmeasurable.

Although all matter is essentially the same it exists in two distinct states: in the form of particles having momentum but no mass and travelling at the invariant speed of light and particles with mass which are at rest relative to the former ones. The momentum of a particle is the product of its mass and velocity and in the final paragraph of the section in the Appendix on Special Relativity the equation

$$E^2 = m_R{}^2c^4 + p^2c^2$$

is derived. Maxwell linked the momentum, p, of an electromagnetic wave:

$$E = pc$$

These two equations can only be mutually true if the mass of an electromagnetic wave (or the photons which constitute it) is zero. This is in accordance with the Special Theory in that if a particle could be gradually converted into energy its velocity would need to increase as its mass decreased since its total energy is fixed at $E = m_Rc^2$ and its kinetic energy $E = ½mv^2$ increases with velocity. As no energy is being fed in from outside E can only stay constant if m diminishes, tending to zero as v tends towards c.

Matter possessing mass can only interact with other mass-possessing matter by absorbing or losing energy in the form of mass-less electromagnetic particles travelling at the speed of light. Let me write this in a slightly different way. Energy transfer between one body and another is effected by particles of energy which have momentum but no mass and which always travel at the invariant speed of light. When one of these particles is absorbed into a body it (the particle) is stilled and the body's mass increases. If the momentum of the energy particle was p the gain in mass will be p/c.

If energy is applied to a particle its mass increases; if the energy causes it to accelerate then as its velocity tends to that of light, its mass tends to infinity. A photon is a conversion of some mass into energy by way of a discrete quantum.

v) Because of this duality measurements which rely on one or other aspect cannot be made simultaneously to arbitrary degrees of accuracy.

Let us take a simple wave such as that caused by dropping a stone in the centre of a pond and for the sake of argument say that the rate of successive peaks passing any given point (that is, the frequency of the wave) varies as the ripple spreads across the pond and that the distance between the peaks (the wavelength) also varies. The only way we can measure the frequency at any position is to time a minimum of two peaks passing the point. As the wavelength continuously varies a precise measurement can only be made at any given time and we have a fundamental difficulty of simultaneously measuring frequency and wavelength: frequency needs time to be measured; wavelength must be measured at an instant of time. However small the variations of wavelength and frequency are there is a limit to the accuracy with which both can be measured at the same time or at the same location. There would be no meaning in trying to tie any precise measurement of frequency to a precise measurement of wavelength. The greater the precision of the measurement of one of the quantities, the less precise must be the measurement of the other, if they are to be tied together.

Apart from this fundamental impossibility of giving precise meaning to simultaneous measurements of frequency and wavelength there is the added complication of any measurement disturbing the object measured because there needs to be an energy exchange, and therefore a transference of mass, between the observer and observed. If the exchange is small in relation to the observed (a few photons to or from a football) the act of measurement has little effect but clearly a photon on a photon makes any measurement meaningless.

In particular, we are unable both to determine the velocity and position of anything to the same degree of accuracy. The better its velocity is measured, the less sure we are of its position. The two are linked in principle by the formula first suggested by Heisenberg: $\Delta x * \Delta v \geq h/m$ where Δx is the accuracy to which we are able to measure a location and Δv the accuracy to which we are able to measure a velocity. For an electron of mass $1*10^{30}$ kg, $\Delta x*\Delta v \geq 1/1500$. That is, if the position of the electron is known with an accuracy of 0.5% (that is, one part in two hundred) then

$(1/200)*\Delta v \geq 6.63*10^{-34}/1*10^{-30}$ giving $\Delta v \approx 0.15$:

that is, the best possible estimate of its velocity could be wrong by some 15%. If the position is known to an accuracy of one part in forty, its velocity could be determined to approximately the same limit.

In the case of a large atom of mass $1*10^{-25}$ kg, $\Delta x*\Delta v \geq 6.6*10^{-9}$ and here measurements accurate to one part in ten thousand can be matched.

In the case of a small ball bearing with $\Delta x*\Delta v \geq 10^{-31}$ accuracies better than one part in a thousand million million can theoretically be simultaneously attained, and Newtonian physics reigns supreme.

Finally, in the case of a photon, with a mass of zero and a known velocity (never mind how accurate) h/m is infinite, and nothing meaningful can be predicted about its position.

CHAPTER 3

KNOWLEDGE OF ABSTRACT THINGS

Thus if I take away from our representation of a body, all that understanding thinks as belonging to it, as substance, force, divisibility, etc., and also whatever belongs to sensation, as impenetrability, hardness, colour etc.; yet there is still something left us from this empirical intuition, namely, extension and shape.[1]

The New Shorter Oxford gives as the primary meaning of Know: "Recognize, perceive; identify; *specifically,* perceive (a thing or person) as identical with something or some one already perceived or considered." (As stated earlier, in its definition of Perceive, the word Know does not appear, so Know is treated as the primary word.) So far I have been writing largely about sense perception and speculating about the nature of a conscious perceiver and the interrelation between the perceiver and perceived. I have been dealing with the first part of the dictionary definition given above. It is now time to move on to consider the second part, the specific meaning which entails an interrelation between perceptions themselves, leading to a different sort of perception.

The usual examples of abstraction are number and space: five as an abstraction, not five this or that; an octahedron qua octahedron as opposed to a cube. Are these true abstractions? It depends a little on the circumstances. The notions *five* or *cube* entail some images when used in isolation in ordinary speech and only perhaps become truly abstract in mathematics. It is quite easy to write a computer program to simulate a conditioned reflex and 'train' it to count how many discrete objects have been placed in its line of vision. After all, all computers count all the time. Whenever I open a file on mine, the page, line and column numbers are displayed (among others) at the bottom of the screen. Programs can be written to define to the computer what is to be meant by a discrete object, such as a typographical character. Given a page of text it can count these,

1. Kant. Critique of Pure Reason. Transcendental Aesthetic: Introduction (Everyman 1934)

or the number of words, or the number of paragraphs. Given a good video camera link it could count discrete objects but these would need to be carefully defined, and the programming would become very complex. How many objects can I see without moving my head? My computer, a mug, a table with a vase of white roses, a few chairs, a clock..... Is each rose separate object? What about the hands on the clock or the numerals? The mug's handle? Its pattern of coloured capital M's? The small coffee stain on the inside of the rim, the vestigial reflection of a window just to the right of this and the lens of coffee visible towards the bottom of the mug? There is no end to this. Well, I suppose the computer could be programmed to ask for definitions so that it would recognise a mug or a vase of flowers or a clock as single objects but for all the sophistication of the programming I do not believe that it would have any more idea of what "2" means than does my computer when it displays, as it currently does, at the bottom of the screen:

Page 2 Sec 1 2/2 At 9.8cm Ln 16 Col 53 Wed 08.55[2]

What about space? Clearly a computer can be programmed to recognise a square or a pentagon, a cube or an octahedron. These are not really abstractions: they are specific line objects. We cannot imagine a pentagon without visualising five lines but what about space itself? Even in the concrete we cannot easily understand it let alone try to abstract from it.

Yet in the mathematical treatment of a pentagon when it is reduced to x's and y's we can get closer to the notion of abstraction for we can be trained to think naturally in algebraic terms, to deal in and manipulate symbols without concurrent imagination of lines defining shapes. I remember years ago the pleasure derived from working out why the square root of two can never be expressed as a proper fraction[3]. Then I wondered about decimals (this was while lying in bed at night during a worrying and sleepless period in order to try to keep the mind occupied) and formulated two rules.

1) All proper fractions can be expressed as finite decimals (as in the case of a quarter or two fifths) or recurring decimals (for example a seventh or four ninths – not so obvious).

2) All finite decimals can be converted to proper fractions by dividing the digits after the decimal point by ten to the power of the number of digits (by definition) and all recurring decimals dividing the recurring portion by as many nines as there are digits after the decimal point to the point of recurrence.

Examples make this clear:

$0.32 = 32/10^2 = 32/100 = 8/25$ (irreducible)

$0.333 = 3/9 = 1/3$

$0.142857142857142857.... = 142857/999999 = 1/7$

2. It has got this last item wrong by several hours and is quite unaware of it.

3. Assume that the square root of two is a fraction a/b where a and b are integers such that neither a nor b have any common factors. Then neither will a^2/b^2. Therefore the assumption has to be wrong.

The point of this example is that the numbers do appear to be truly abstract – there is no mental image of objects apart from the numerical symbols themselves. We have an abstract rule which was true before it was 'discovered'. If it was true it must have existed somewhere. What was the form, the essence of its existence? Plato was, I think, the first to articulate the problem clearly and the answer most unclearly: he implies that there exists at some deeper level than this miserable material world a world of forms of which we only perceive shadows. It does not take much consideration to see that such a belief leads to an infinity of forms and a totally impossible situation, and yet the problem persists to the present day with an eminent mathematician agonising over it:

> *The world that we know most directly is the* ***world of our conscious perceptions****, yet it is the world we know least about in any kind of precise scientific terms. That world contains happiness and pain and the perception of colours... It is a world containing mental images of chairs and tables, and where smells and sounds and sensations of all kinds intermingle with our thoughts and our decisions to act.*
>
> *There are two other worlds that we are also cognisant of – less directly than the world of our perceptions – but which we know quite a lot about. One of the worlds is the world we call the* ***physical world****. It contains actual chairs and tables... electrons and photons and space-time. It is not at all clear why the world of our perceptions should have anything to do with the physical world, but apparently it does.*
>
> *There is also one other world, though many find difficulty in accepting its actual existence: it is the* ***Platonic world of mathematical forms****. There, we find the natural numbers, 0, 1, 2, 3,....., and the algebra of complex numbers.... We find Euclidean geometry's Pythagorean theorem... and geometries other than the Euclidean one.... Also in this world are Maxwell's electromagnetic equations as well as Einstein's gravitational equations...*
>
> *What right have we to say that the Platonic world is actually a 'world', that can 'exist' in the same kind of sense that the other two worlds exist? It may well seem to the reader to be just a rag-bag of abstract concepts that mathematicians have come up with from time to time. Yet its existence rests on the profound, timeless and universal nature of these concepts, and on the fact that their laws are independent of those who discover them.*[4]

Now, I believe it is a misuse of language to say what we know most directly are our own perceptions[5] since it is through our perception that we know a chair. To know our perceptions would need a knower apart from them, an I within me which would lead to an infinite regression: if we could perceive our perceptions we could perceive our perceiving our perceptions... Perceiving is an act of consciousness; it might involve an external or an internal stimulus.

---

4. Roger Penrose, Shadows of the Mind OUP 1994 8.7
5. cf Wittgenstein, almost at the end of his Philosophical Investigations: 'I can know what some one else is thinking, not what I am thinking. It is correct to say "I know what you are thinking", and wrong to say "I know what I am thinking." (A whole cloud of philosophy condensed into a drop of grammar.)' II xi - Blackwell 1963 222[e]

External stimuli such as light or sound involve the exchange of photons and electrons in our sense receptors, which then travel via a series of nerve fibres to our brains, when we experience the sensation of seeing or hearing and where some form of permanent or semi-permanent data base is established.

When we remember a thing (whether it be a chair or that two and two make four) an internal stimulus – sometimes voluntary and sometimes involuntary – "switches on" some part of the data base and we become conscious of an image or a thought. This switching on also involves the absorption of photons by electrons within our brain cells. Apart from a tentative mapping of locations within the brain and a certain amount of speculation, almost nothing is known of the nature of the data base or the mechanism by which parts of it are awoken into consciousness. The perception in all cases is the becoming conscious; the database is something quite separate, although whenever our perception involves the recognition of the thing perceived some recourse to the data base is necessary. We cannot perceive a perception.

Each of us indivisibly perceives and one of the things we can perceive is a chair and another is that two and two make four. Is the latter an abstraction from sense perception?

Can we perceive memories? Can we become aware that we are aware? Aware that we are aware that we are aware….?

We perceive our memories only in the sense that there has to be some activation of that part of the brain, those brain cells, where memories are stored. We cannot look at the cells in any sense as we look at an external physical object. I, the whole of me from my toenails to the bald patch on the top of my head, look at a chair; I, the whole of me try to remember what a particular chair looks like. The mechanism of seeing is more or less understood as far as the signals from the object seen travel to the brain; their passage within the brain is very obscure; their entering into consciousness a complete enigma. The mechanism of trying to remember and remembering is also wholly obscure. It is a simpler, and for all practical purposes in everyday use, a more accurate expression to say, not that we perceive a memory, but that we remember.

The Platonic world does not rest on the profound, timeless and universal nature of its concepts but on the nature of space-time-matter. This may well be profound (depending on the meaning we give the word) but it is certainly not timeless, nor, strictly speaking, is it universal. Truly, as Augustine said, there was no time before time and there is no space outside space or space without matter and with no time or space or matter maths has no existence or meaning. Given a different form of space-time-matter an entirely different form of maths is easily imaginable (say, one in which one and one sometimes make two and sometimes three) and in this sense it is not universal.

All knowledge, of its very nature is abstract, and as for the knowledge of persons the difference in knowing is essentially in the known rather than in the knowing. It is possible to imagine a computer being programmed to have a similar database in its memory and to be able to recognise objects, whether concrete or abstract, with which it is presented. There are two other phenomena I would like to discuss where something quite separate might obtain and where it is difficult even in principle to imagine that they are computable: the notion of insight and the discernment of truth itself.

INSIGHT

Let me give three examples of insights. In the first it is easy to trace the path that a computer might go through, mimicking what goes on unconsciously in the brain. In the second a computer would most likely arrive at the right answer with no element of insight and in the third I am at a total loss.

The first example is the solving of a crossword clue, published in The Times many years ago:

Oiled Silk? Stuff! (5,2,1,5)

The computer (I imagine) would consult its dictionary and phrase book and find:

Oiled, (8th definition) See Well-oiled
Well-oiled : Drunk.
Silk (6b) *Informal* a Queen's Counsel
Counsel (4) a barrister Member of judiciary
Judiciary (1) …pertaining to judges
Stuff (8) Nonsense.
Not oiled: Sober
Sober……..judges
Sober as a judge

It will of course have to compare each of the meanings given in its dictionary of *Oiled* and *Silk* and *Stuff* with each other (several thousand comparisons) and check each of these with its phrase book until a suitable one is found.

And I suppose a similar process goes on in the solver's sub-conscious: it is essentially a mechanistic process, an algorithmic one, following a set of clearly defined rules. As has previously been said it has a perfectly satisfactory evolutionary explanation, including the feeling of satisfaction when a compelling connection is made.

The second example is the well known puzzle devised by Edward de Bono: Take a glass of vinegar and a glass of milk. Put a teaspoonful of vinegar into the milk and stir vigorously. Take a teaspoonful of the mixture and put it back into the vinegar. Which will be the most polluted?

A computer would calculate the percentage of vinegar in the milk in terms of the relative volumes of the glass and teaspoon, even if these are not known. It would then compute the quantity of milk in the vinegar after a teaspoonful of the mixture has been returned, and produce an answer. If we try the same process we get into a fearful muddle and there is a much simpler approach to the problem.

Since the final volume of milk and milk mixture is the same as the initial volume of milk – and the same for the vinegar – the milk must have lost exactly the same quantity as the vinegar it has gained, and this quantity has been transferred to the vinegar. They will be equally polluted and no calculation is needed. A simpler example would be to take a bowl of one hundred red beads and another with a hundred blue ones. Take twenty beads from the red bowl, put the in the blue one and stir them. Take twenty beads at random from the mixture and put them into the red bowl. Each bowl now has the number of beads that it started with. However many blue beads the red bowl has gained must equal the number of reds that the blue bowl has gained. In this case the algorithm is straightforward:

After first transfer red bowl has 80 red beads and blue bowl 100 blue and 20 red. After mixing and transferring back x red and y blue, with x + y = 20. So the red bowl will have 100 – x blue beads.

The blue bowl must have x red beads as there is nowhere else they can have gone!

Without the need to calculate percentages a computer could count atoms of milk and vinegar and use the same algorithm, so no mysterious insight is needed.

For the third example, look at the definition in symbolic logic, using Whitehead and Russell's notation:

p ⊃ q. = .~p v q Df.[6]

We are interested solely with an understanding of the definition, of what is meant by the various symbols.

The letters p and q stand for propositions such as "This is a cow".

The symbol ⊃ means that p infers q: "If this is a cow it will have horns".

The dots merely isolate the various parts of the definition: they act as brackets.

= has its ordinary meaning.

~ is a negation: ~p in this instance means "It is not a cow".

v means either or, so ~p v q means "Either it is not a cow or it has horns".

Df means that what has gone before is a definition.

Even though it is the proposition which is being defined the implication is that, "If this is a cow it will have horns" is obviously the same as saying, "Either this is not a cow or it will have horns, as this is the point of the definitions.

Now, when I first read the definition in the Principia Mathematica it was not immediately obvious to me that, "If this is a cow it will have horns" was equivalent to, "Either this is not a cow or it has horns" and it needed some reflection to see that it was, after all, obvious and even sometimes used in common speech. Imagine trying to work out the reason for a water leak in the cooling system of a car and coming to the conclusion that it was due to a broken head gasket. While the cylinder head is being removed you could say to the mechanic, "Well, either I am wrong or the gasket will be broken, rather than, "If I am right the gasket will be broken".

Here it seems to me we are much thornier ground and whether such an insight is even in principle computable will depend on where we get our understanding of logic from.

---

6. Principia Mathematica Chapter 1 Definitions Cambridge 1910

There are at least two possibilities: it is an evolutionary product or it is something that cannot be explained by evolution. It can be (and commonly is) argued that our logical thought is due to the 'hard wiring' of our brains and that this is purely a function of our evolution. The great difficulty with this thought is that we are ultimately left with a situation in which a conglomeration of atoms, however marvellously organised, are able to form judgements of what is true. That the organisation, the 'hard wiring' of our brains could be purely the result of evolution is entirely possible; that it is necessary for us to be able to reason is entirely reasonable; that it is the whole truth is very unsatisfactory.

But then, since we have no control of insights – we can strive and strive to grasp a point and fail for many days and then gain understanding when we are going about our normal everyday tasks and thinking of something quite different – it is difficult to see how it can be argued in these two examples that this is the result of some mysterious spiritual component within us. A supposed spiritual element within us is commonly identified with the intellect and free will, yet neither have any control in the matter of insights! Clearly when we have an insight some connection is being made. A search function of which we have been unaware has been going on in our brains. The criteria for the search is probably conditioned by the intensity and breadth of the initial enquiry but could potentially cover the entire huge database which every human carries. It seems likely that the length of time the search will be continued will be determined by other demands on the brain but this is mere speculation. It is a commonplace that unsolved crossword clues in the evening are sometimes very simple in the morning, although if not re-attempted in the morning it is unlikely that the answer will be ever come to mind. It is also a commonplace that in matters we have agonised over in an attempt to understand the answer can pop up unbidden several days later. As was stated in the opening chapter all this indicates that some form of 'mechanical' process is going on, and one that could, in principle, be carried out by a well-programmed computer but whether any computer can be self-judging is an entirely different question.

CHAPTER 4

BOOK KNOWLEDGE

It must be true: I read it in the newspaper.

I beseech ye, Brethren, in the bowels of Christ, always consider that ye might be mistaken[1]

There is a huge mass of things we know of which we have no first hand experience; there is an even larger mass of which we claim knowledge when in truth we have no real opinion at all. This chapter is an attempt to clarify the extent and limitations of our knowledge of things which are outside our own direct experience.

There are those of which I have no first hand knowledge at all although I have seen plenty of pictures of them, including television: I know that Washington is the capital of the United States.

I have never been to Washington but I am no less sure of its existence than I am of that of London or of it being the seat of government. The sheer weight of uncontested evidence makes it inconceivable that this should not be the case.

Then there are those of which we have some first hand knowledge but not of the actual point we are asserting: I know that London is the capital of England.

I lived in London for several years and have visited it hundreds of times. I have seen the Houses of Parliament on countless occasions though I have never been inside them. There is no room for reasonable doubt that they are the Houses of Parliament, the seat of government, and beyond imagining that some vast conspiracy exists leading us to believe that London is the capital whereas in truth Morton-in-the-Marsh is. I know beyond doubt that London is the capital city.

1. Oliver Cromwell, letter to Synod of Church of Scotland, 5th August 1650

Much as in the case of London, I know that Birmingham is England's second city. I know that this is true but I also know that some ignorant people claim the title for some other city, such as Manchester.

These examples are trivial and to some extent a matter of definition.

More seriously is the difficulty of sorting out from the mass of scientific dogma those things, if any, which are incontestably true, those which are almost certainly true and those over which some or great doubt exists. In the latter categories three examples will be considered, evolution, global warming and genetic modification.

The things which are incontestably true are facts of our own observation and by extension those where, as in the case of Washington, there is such a mass of corroborative writing by diverse people, photographs, news items, accounts of friends who have been there and so on that it is impossible to doubt its existence.

No scientific theory is incontestably true but they do range from the almost certain to the highly speculative. For instance, it is impossible to doubt the truth of Galileo's observation that if two stones are dropped from a height, one being twice as heavy as the other, they will fall at much the same speed. Even if we have not carried out the experiment ourselves (and I do not remember doing so) it is such a simple thing to do, the line of a priori reasoning so persuasive and the universal acceptance so well documented that there is no room for any doubt.

A modification of Galileo's reputed a priori reasoning is as follows: imagine two equal weights being dropped from a tower side by side, but not touching. There is no reason why they should not keep pace one with the other. Now let them drift together. We now have an object of double the weight but there is no reason to think that this will suddenly speed up. A simple experiment dropping two unequal weights is necessary to confirm this intuition and it proves correct. Whether they will fall at exactly the same speed and whether bodies of different densities all behave identically are further matters? Simple experiments quickly establish that lumps of similar masses of iron and aluminium do indeed keep pace with each other. Do they have exactly the same accelerations? Here an element of doubt creeps in. We know that air resistance does affect the outcome. If the experiment is carried out in a vacuum would we find that objects of unequal mass and unequal density fall at identical rates of acceleration? I have no doubt (I know from experience, not of this but of many other experiments) that the results of any actual experiment will show a slight scatter of accelerations even for the same object being dropped many times from the same height, let alone different masses falling different distances. Ah, I would say, absolute vacuums are impossible: there will always be the odd molecule floating around because the evacuating machinery was not one hundred per cent efficient and some may even flake off from the container itself; absence of effect from radiation which will affect acceleration cannot be achieved: cosmic rays will penetrate any container. An abstraction from the results is always necessary to propound the law: *gravity affects all masses equally.* The experiment has been done with great precision, down to an atomic scale, and the theory confirmed to an accuracy of better than two parts per billion.[2]

Without wishing to go into complexities of understanding of space, there is for me no shadow of doubt that the earth is roughly spherical in the sense that a tennis ball is spherical. The accumulated experience of countless thousands of travellers and navigators, even before the more recent evidence from satellite observations, put the matter beyond doubt. Despite this, the International Flat Earth Society still exists, and some of its members, as educated and intelligent as us, doubtless sincerely believe, and make a reasoned case for believing, that the earth is flat.

Newton's Laws, Relativity, Quantum Theory, current cosmological theories—what about these? With respect to the first three although I believe (I have to say believe as I have very little first hand knowledge) that the enormous mass of corroborative evidence and the lack of any adverse evidence provide a substantial argument in favour of their correctness, we know that some of Newton's fundamental postulates were erroneous and that both Relativity and Quantum Theories lead, to say the least, to great obscurities of interpretation; in the case of the former also to certain unresolved anomalies. My judgement is that although I am scarcely in a position to judge, it is probable that these fundamental theories will be replaced by successively simpler syntheses in the fullness of time, requiring insights as radical as those of Copernicus and Einstein in their day. I do not believe that they will be shown to be false in the way that the Ptolemaic solution has been but it should not be forgotten that in its extended day this was revered as near dogma by a large body of the most educated men of their time, also no less intelligent than us.

Ptolemy's system is wrong primarily because its predictions do not accord with observation and secondarily because it relies on a series of ad hoc assumptions to make it work at all. Firstly there is the assumption that all heavenly bodies move in perfect circles round a static earth: well, we observe this day in and day out (except when it is cloudy) so this is not entirely an ad hoc assumption and as far as the stars are concerned it works pretty well. Some refinement is needed because with the exception of the Pole Star which stays pretty still, only wobbling two or three degrees in the course of a year, although the remainder still wheel in great circular arcs on any given night, the arcs themselves move in position throughout the year. When it comes to the planets, these too move in circular arcs on any given night, but these arcs move all over the place not only during a year but during centuries. By placing the earth off centre and by proposing that the planets moved in epicycles, also sometimes eccentric, Ptolemy was able to publish tables which fairly closely predicted the positions of the stars and planets. He took much of his information from earlier astronomers and

2. Peters Chung and Chu, Stanford, reported in Nature August 1999

it has been recently claimed that in some cases where observation was at odds with his theory he falsified the data and committed a crime against science.[3]

What is true for fundamental physics is also true for all other branches of science, particularly those which are not susceptible to direct experiment, such as the theory of evolution. It is known by plant breeding, for example, that species are not immutable and that some evolution in general has occurred is beyond reasonable doubt but that life evolved from some primordial swamp, well, it is not beyond the bounds of imagination that this is simply wrong, and that at least some creationist theories are right. None of the mechanisms, from Dawkins blind chance and Darwin's survival of the fittest to the teleology of de Chardin are likely to be the final word on the matter. Teilhard de Chardin, in The Phenomenon of Man, (Editions du Seuil 1955,Collins, 1959 – I bought my own copy two months short of fifty years ago) puts forward the idea that matter has a propensity to unite in ever more complex stable substances, starting with elementary particles and finishing, to date, through evolution, with man. Two eminent scientists had very different contemporary opinions on the work. Julian Huxley in his opening words to his introduction to the English edition describes it as a very remarkable work by a very remarkable human being; Richard Dawkins quotes Peter Medawar's criticism that the work "resorts to that tipsy euphoristic prose poetry which is one of the more tiresome manifestations of the French spirit." He adds that although he was himself captivated when he read it as an over-romantic undergraduate he now, like Medewar, regards it as the quintessence of bad poetic science.[4] Richard Dawkins thinks that the ability to form stable combinations is a property of matter that has happened not strictly by chance but by the inevitable outcome of vast numbers of chance mutations over aeons and that therefore increasing complexification is inevitable given enough matter and enough time. Qua evolution there is not all that much between the two views but not only does de Chardin differ in his views of what happened before there was matter to evolve but he also believes that evolution is leading to a union of God and mankind through and in Jesus Christ, and here of course Dawkins and he emphatically part company.

Other theories, or refinements of theory will doubtless emerge and full understanding is unlikely to be at either extreme of creationism or blind chance. The various branches of creationism, themselves vitriolically divided, rely on a number of literal interpretations of Genesis and, to a lesser extent for Muslims, on the Koran. Their writings are, in my experience, disreputable in the sense that they are extremely eclectic and almost totally lacking in meaningful references of the hotchpotch of quotations they assemble. A good example is the Jehova Witnesses' tract, 'Is There a Creator Who Cares About You?' (Watch Tower

3. P R Newton The Crime of Claudius Ptolemy Baltimore MD 1977. Quoted in J J O'Connor and E F Robertson's excellent article in the Ptolemy article in the Mac Tutor Series published by St Andrews University Maths Department on the web: www.gap.st-and.mcs.ac.uk/~history/Mathematicians/Ptolemy

4. Loc cit

1998). Although it is possible that the creationist case it is arguing is in fact correct, the method of argument is intellectually dishonest and does a disservice to its cause. It is dishonest in its mass of undocumented quotations, many of which could well be out of context, in its failure to put the counter arguments, in its assumption that the current state of knowledge will be final (what has not yet been demonstrated will therefore never be demonstrated) and most seriously in treating the Bible uncritically as being of the same nature as scientific research. Having said this, it is not nearly as extreme as earlier versions of the same tract. There is still a mass of literature claiming that the universe was created four thousand years ago, and citing discrepancies in carbon dating to cast doubt on the deduction that the earth is many millions of years old. It entirely ignores the huge mass of corroborative evidence from other sources. Dendrochronology (tree ring dating), with an ability now to date approaching ten thousand years, geology, geography, history itself, all come together to provide a truly compelling case. Although it seems to me just possible that the earth is not many millions of years old, it is certainly many thousands.

As an aside, I think The Blind Watchmaker also exhibits some of the traits of the Creationist literature. The opening passage of Peter Atkins *the Creation*[5] is quoted and the gloss added "*...rather that he is satisfied that biologists can explain elephants, provided they are allowed to take certain facts of physics for granted. His task as a physical scientist, therefore, is to justify taking those facts for granted. This he succeeds in doing.*" The fact in question is the eruption of the universe by a quantum quirk from absolutely nothing – no space, no time, no matter, no energy: nothing. The fact is, and Dawkins make no reference to this, is that Atkins far from presenting spontaneous creation as a fact merely proposes it as a satisfying speculation.

The most powerful argument in favour of the likelihood of evolution stretching back in time to the big bang (if there was such a thing) is that there are no clear boundaries between small molecules – stable groups of atoms – such as water or methane, or huge ones, such as an enzyme, or single cell plants or animalcula; there are no clear boundaries between single cell and multi-celled animals or between any species before they begin to diverge. That species have become, and are becoming, extinct can only be denied given some extreme theory such as that proposed by Philip Henry Gosse in his nineteenth century work Omphalos: that God created the fossil record to snare the unbeliever. His son, Edmund, in his early autobiography, Father and Son, vividly describes the religious background of the Plymouth Brethren that led his father, an eminent zoologist, to his preposterous theory. The father anticipated fame and fortune from the work and reaped ridicule. The notion of a fraudulent God has inherent self contradictions that render it absurd and yet Christians believe that His Son taught us to pray, Lead us not into temptation. That species can and do mutate is uncontestable: by genetic modification we are able to engineer this in the laboratory.

---

5. op cit

The belief or otherwise in this or that theory of the mechanics of evolution has no great practical effect on our lives at least in so far as we are able to change its course. That it has had a huge effect on our understanding of our existence in general, and of many religious beliefs in particular, is undeniable. The universal belief in the literal truth of both Old and New Testaments has become untenable and the ongoing reappraisal and critical exegesis of these works is leading to fuller and deeper understanding of religious beliefs. (As far as the Koran is concerned this revolution is only just making a hesitant start.)

Global warming is a much less sure phenomenon but paradoxically has a far greater practical importance. There is no doubt of the truth that global warming is currently happening at least in the short term and understanding of its causes will make possible schemes for reversing or slowing down the process which could have a huge impact, probably not on our own livelihoods but on those of future generations. My striving to understand fundamental scientific theories is therefore not mere self indulgence. Whereas it is true that it is of little or no practical significance to any one but himself whether a person does or does not grasp some mathematical point there might be matters of great importance on which we have an obligation to have an informed opinion in order to be able to act in a responsible manner. In the case of global warming I am sure that the great majority of us (emphatically including myself) can scarcely have any worthwhile opinion at all as to the actuality in the long term – over a period of a few centuries – or as to the present causes.

I know from my own experience that sixty years ago in Hertfordshire we were able to skate outdoors every other year or so, and forty years ago one was assured of snow at Glencoe at New Year and that neither of these is now the case. I also know (though obviously not from my own experience) that there have been times both in the recent and distant past when similar or even warmer conditions have prevailed in Britain and (although I am not so sure about this) in the world as a whole and times when much colder conditions have prevailed. The reasons for these cyclical changes are at best partially understood and it is, to say the least, possible that the current theories (changes in the tilt of the earth's axis, release of dust into the atmosphere due to meteorite impact or to volcanic eruption, complex changes in composition of the atmosphere leading to greater absorption of both direct radiation from the sun and reflected radiation from the earth . . . .) which explain some of the apparently sudden marked changes in global temperature which have occurred are not only inadequate but might even be simply wrong in some respects. I have read that a large body of scientists believe that carbon dioxide, methane and nitrous oxide are chiefly responsible for the greenhouse effect and that chlorofluorocarbons (CFC's) are destroying the ozone layer in the upper atmosphere which prevents some direct radiation from the sun reaching the earth. I am not wholly convinced of the truth of these theories but that does not seem to be of practical importance since it seems plainly sensible to reduce emissions of these gases in any case.

I am a little sceptical of consensuses of experts both because they have so often historically been wrong and because research grants inevitably cause some conflict of interest. I have read and heard that we, mankind, are now responsible for appreciably altering the proportions in which the gases are present in the atmosphere and (much as in the case of genetic modification) unless we are convinced that we are not causing harm we should be extremely wary of altering natural balances. Atmospheric pollution, as in the case of dumping at sea, is no longer a case of merely a drop in the ocean—indeed, the former is now expected to cause a rise in the latter! Have I any evidence for even this last assumption? Have I good reason for saying that we should be wary of upsetting natural balances? Am I right to abrogate any responsibility for taking any action in these matters? Is it my duty to inform myself a littler better so that I am in a position where I am able to start to make a reasonable judgement? Is there any end to this vein of questioning?

Let us move on to another matter in which the inadequacy of book knowledge is most obviously illustrated. It is one in which I have considerable academic knowledge gained over many years. I know all about sex: I have read widely on the subject.

CHAPTER 5

KNOWING PERSONS

Where , like a pillow on a bed,
A pregnant bank swelled up, to rest
The violet's reclining head,
Sat we two, one another's best;

Our hands were firmly cemented
With a fast balm, which thence did spring,
Our eye-beams twisted, and did thread
Our eyes upon one double string;

So to' intergraft our hands, as yet
Was all our means to make us one,
And pictures in our eyes to get
Was all our propagation.

As 'twixt two equal armies, Fate
Suspends uncertain victory,
Our souls, (which to advance their state,
Were gone out), hung 'twixt her and me.

And whilst our souls negotiate there,
We like sepulchral statues lay;
All day, the same our postures were,
And we said nothing all the day.

This ecstasy doth unperplex
(We said) and tell us what we love,
We see by this, it was not sex,
We see, we saw not what did move

When love, with one another so
Interanimates two souls,
That abler soul, which thence doth flow,
Defects of loneliness controls.

We then, who are this new soul, know,
Of what we are composed, and made,
For, th'atomies of which we grow,
Are souls, whom no change can invade.[1]

---

1. From The Ecstasy, John Donne.

KNOWLEDGE OF OTHERS

In the first two chapters I tried to examine some fundamental aspects of knowledge of the material world through sense perception which is the only means I have (at least in so far as I have any experience) for gaining any knowledge of any sort. Matter, time and space in so far as they are perceived were seen to be in part products of the conscious perceiver and the possibility was raised that until perceived they did not realise the fullness of their existence. Is there then anything essentially different in my perceiving a stone and in my perceiving you or perceiving myself? Can there be a meeting of minds? Do I mean by this, can there be a meeting of souls? And if I do not mean that, then what do I mean by a meeting of minds?

The distinction, if any, between mind and soul will be explored later but for the moment we will assume that the notion of soul encompasses some wholly immaterial element. A meeting purely of souls would therefore entail some form of extrasensory perception, that is, perception which has no material element in it. Whilst certainly not ruling out the possibility of this phenomenon, it is not the subject of this chapter. By meeting of minds I mean the mutual perception of one intelligent being by another rather than the everyday usage of like-mindedness.

When we say we know a stone, some elements of what we mean by the word "know" in this sense are dealt with in the first chapter. The purpose of this chapter is to explore some elements of the meaning of "know" when I say I know you. The examples I gave in the Introduction were

I know Mr Smith	*I know him to say "Good morning I know him intimately; he is one of my best friends. In the former he is being treated more or less like a tree. In the latter has something quite separate crept in?*

Knowledge of a stone consists in a number of things such as the memory of the perception of a particular stone, understanding its crystallography or chemistry or geological significance and so on, and my knowledge of you consists at least in part of the same. "I know a stone", really means that I know about a stone or I have a memory of a particular stone. When I say I know you, do I mean more than that your image is in my memory and that I know about you?

Scientific knowledge allows us to predict reactions. A theory is held to be true if events that should happen according to it actually do happen in every single trial provided nothing else intervenes. Knowledge of another person (or even of ourselves) is never of this nature and if we truly possess some freedom can never be of this nature. There is, of course, a great deal that we can predict.
The whole of medical science (just as of veterinary practice, just as of the work of a good car mechanic) depends on an ever-increasing understanding of the

workings of the human body, down to the level of molecular interactions and beyond and to an increasing extent is reliable. Although there is still an unfathomable extent that remains to be understood, it does in very many cases make possible reliable diagnosis and treatment of a range of diseases. If the crown of science is the ability to predict, and the concomitant of this, to be able to manipulate, then incontrovertibly human biology is at present a more successful science than psychology. The concomitant of human biology is medicine. By doing *this* we will allow or prevent *that.* Psychology's concomitants, psychotherapy and to an even greater extent psychiatry are at present far less certain and it might be thought that this can be used as an argument for there being some extra ingredient in man. This is not the case, for precisely the same arguments can be used concerning animals – just consider a highly strung racehorse – and, although psychology is not quite the right word, to plants where genetic modification is a very real science but the results in particular plants by no means certain.

Yet for all the advancement of medicine and psychiatry we are no less surprised than were our forebears by those we know most intimately, husbands, wives, lovers and perhaps most of all our own children, whom we have observed from the moment of birth. It is a commonplace, "I thought I knew him". We are occasionally hugely surprised by our own actions: "I surprised myself", is also a commonplace.

The more we consider the matter, the more it becomes clear that when we say we know another what we mean is that we recognize them—that they are in our memories in many moods and situations. Generally speaking self-knowledge and knowledge of inanimate and animate beings is all of the same nature: the difference is not in the knowing but in the knower or the known. The only difference in knowing a stone and knowing a dog and knowing myself and knowing you is in the objects known.

What then of the experience of John Donne in his Ecstasy? He says in the poem that for all that their hands were firmly cemented, intergrafted, all the day, it was not a sensual experience but a true meeting of souls. This form of expression is more common in mystical writers than erotic ones and there is no hint, I think, of this sort of experience in either his marvellous later sermons or in the Holy Sonnets and other Divine Poems. I hope to write a little on mysticism later and in the meantime can only repeat that it is outside my personal experience and while by no means denying its possibility I am not at all sure that there is not a straightforward scientific explanation for what is being claimed in most cases, at least in so far as human beings are susceptible to scientific explanation –just as I believe that most of what is claimed as extrasensory perception is, strictly speaking, nothing of the sort. The great gamut of phenomena ranging from déjà vu and previous existence experiences through telepathy to fortune telling, bending of spoons and extra terrestrial visitations are on the whole, and quite possibly entirely, either hocus pocus or delusion or coincidence. For most of them there are perfectly reasonable explanations and Ockham may be invoked. As far as John Donne's Ecstasy is concerned it seems to me likely that this form of heightened awareness is in fact completely sensual by which I mean that a full understanding of the nervous system would lead to a comprehension of this out of body experience: it is neither necessarily an illusion nor a delusion but rather an example of the inadequacy of language.

If my knowledge of you is gained only through sense perception in exactly the same way as all my other knowledge (and I have not one shred of personal experience to suggest the contrary) what reason is there to posit anything spiritual in man?

When I started to write this chapter I thought that there was an essential difference in the meaning of "know" in "I know a stone" and "I know Mr Smith" but this does not appear to be tenable.

SELF KNOWLEDGE

O wad some Pow'r the giftie gie us . . . .

Self-knowledge is also gained through sense perception. Without sense perception I cannot even utter Descartes' Cogito. The very act of awareness requires cerebral activity and even were it possible to be aware of nothing the brain would still need to be whirring or the awareness would pass into unconsciousness; if all activity in the brain ceased, would we then we would cease to be persons and disintegrate? Is the act of awareness itself then merely a product of cerebral activity? Is there any sense in saying that I am aware that I am aware? Does this add anything to the simple statement that I am aware? I think there is some sense in saying that I exist as a perceiver as well as a self-perceived and that I know that I exist as both. There are limits of course to my perception of myself placed by the location of my sense organs and to a lesser extent by the inflexibility of my body. I need to tape-record my voice to hear it properly and I need a mirror to see the back of my head. I look at my fingers in much the same way that I look at yours but I feel my headache in a totally different way than I feel yours--in fact I do not feel yours at all. It is quite obvious that whatever disturbance in the brain cells there may be, my headache is only truly a headache when I am feeling it.

I have used the first person so far as the arguments will be entirely from my own experience and will, I suppose, be more or less convincing as they coincide with the experience of the reader. Extrasensory perception or supernatural phenomena may well exist but I have no knowledge of them from personal experience. I have seen no visions, witnessed no miraculous cures, conversed with no angels, managed not even the smallest levitation. Yet, at times I have had intimations of the existence of immensity but I will base no argument on this either. If this experience is not universal I have a suspicion that the suspicion that such an experience might be there for the taking is universal.[2] If nothing

2. In Shadows of the Mind (OUP 1994 1.3) Roger Penrose rejects the view that, "Awareness cannot be explained by physical, computational, or any other scientific terms" and

else, the knowing that I know defines to some extent the meaning of awareness about which I am writing. Does it imply the possession of something beyond mere cerebral activity?

The examination of this age-old puzzle will be considered from two quite different starting points.

The nature of consciousness—conscious that it is conscious—is the first of the two starting points. The operation of free will will be the second. With respect to the first, a totally materialistic theory of evolution leads to some anomalies.

All life has the huge majority of genetic material in common and it is thought very likely that with rare exceptions all living things spring from a common ancestor. Evolution has occurred step by step through minute hereditary changes from ancestors in common.

There is no change in principle in status of any given parent and its offspring: if the parent is classified as a bacterium so must the child be; insects will always give birth to insects and mammals to mammals: individual genetic differences are almost imperceptible. It is only through thousands of generations that first species and then orders diverge.

What applies to the whole obviously applies to its attributes such as intelligence, consciousness and freedom: there can be no difference in principle between the consciousness of a plant, a bacterium or a horse but only differences in degree. This would be clearly seen if we were able to know the genome of say a horse and to trace the changes generation by generation back to a primordial cell: there would be no instance at which we would be able to say of a single offspring *This is a horse* and of its parent *This is not a horse* or *This is an animal* and *This is not an animal*. There would be no point in the evolutionary chain where it could be said that the child could feel pain but not the parent, no instance where an immediate descendent could be said to be conscious and its progenitor not to have consciousness.

It does not require much reflection to understand the force of this argument. There is no need to stop in the present thought experiment at the primordial cell, whatever this may be. It too will have evolved by some or other mechanism (of which currently no coherent theory exists) and the instance at which a non-living cell became a living one will not be possible to pin down.

Step by step, changes in molecular structure, sometimes just of a single electron, sometimes of a chance combination of two molecules following an accidental collision, could be traced back through amino acids to simpler substances and

---

argues only that "Appropriate physical action of the brain evokes awareness but this physical action cannot even be properly simulated computationally". He goes on to say that some ingredient of the former view seems to be involved in the acceptance of religious doctrine. He continues, "If you are a reader who strongly believes that this view, — the scientific road—and try to perceive where I believe this road must ultimately be taking us."

from these via atoms to the big bang. At no stage whatsoever could life or consciousness or intelligence or freedom be said to suddenly appear.

It must follow that all matter is in some real sense living, conscious, intelligent and free, in so far as these are the characteristics of any living creature: if a cat is conscious so is a cabbage, if an ape is intelligent so is an acorn. This is incontrovertible if an untrammelled theory of evolution is true.

What applies to living creatures applies a fortiori to computers. The laptop on which I am typing this is enormously more sophisticated than the first one I owned only thirty or so years ago.[3] Yet with regard to self awareness neither of these computers has more (or less) than a stone and no increase in the speed of operation, size of memory or the amount stored in its memory, its ability to be networked with other computers or run in parallel with them or to form neural nets or whatever will increase their self awareness one iota. It appears to be a change of order and not of complexity and one that proponents of artificial intelligence have not adequately addressed. At no time will the addition of a single critical electronic component make an object with all the awareness of a lump of clay become instantly truly self aware.

That would be clearly absurd: proponents of Artificial Intelligence cannot escape from the thought that self awareness is only a matter of degree, starting at elementary particles and progressing through atoms and molecules to viruses, bacteria and so on through the evolutionary chain until the development of the full nervous systems found in animals, including man. In this scheme the electronic computer would rank perhaps alongside a virus. I am not speaking about the ability to perform complex calculations at phenomenal speeds—or rather, simple calculations at phenomenal speeds since complex calculations can always be split up into simple ones—but about self awareness.

My problem is that using these irreducible elements I cannot imagine any assemblage that would result in self conscious awareness. I said almost total incomprehension because there is one possibility that I can see whereby an artefact could be made capable of being self aware. If it were possible to construct a human being in the laboratory—or even just to clone one (and this is almost certainly possible) would this being be truly human? There is no doubt that the cloned sheep is really a sheep. My view, for what it is worth, is that the android would become human in the same way as any other human embryo: its capacity for true self awareness would be fulfilled. There remains, of course, the alternative possibility, that although the creature would exhibit all the signs of being truly self aware, in fact it was not although I must say that this notion appears to me to be pretty ridiculous.

The question of whether self awareness in a creature implies the—for want of a better word—possession of something beyond the matter out of which it is made still has to be faced and here again I am faced with something close to entire incomprehension. For the moment let us just stick to the sheep. Neither it,

---

3. It was a Hewlett Packard, not much larger than a laptop, with an integrated screen and printer and a memory of 16K.

nor any other animal I have ever observed appears to be self consciously aware in the sense that I am using the words. Here I know I am on dangerous ground, but it does appear to me that where there is possibly an uninterrupted gradation of consciousness from a stone to an elephant[4], for all the persuasiveness of modern anthropomorphic sentiment there is a decisive difference in man. I think it should also be said that this is still the view of the great majority of mankind and has been the view of no less observant man living on the whole much more closely with animals than we do, for countless thousands of years.

Secondly there is the puzzle of how certain aspects of human activity can be governed by the laws of physics and in particular the operation of free will. At one and the same time I find this the more powerful argument and the harder to demonstrate. It is more persuasive because we are all convinced that we do have real freedom and because it is positive; the first starting point is negative in that my inability to understand might just be due to a lack of imagination on my part—this has often been the case in the past. The operation of freedom of will is at the same time perfectly obvious and devilishly difficult to demonstrate because it tends to dissolve and fragment on close introspection and we so often observe others acting absolutely predictably when they have thought themselves to be acting completely freely.

I know (I really do, I really know, I know without a shadow of doubt, I have tried to doubt but cannot) that I have real freedom but I do not know *in any particular instance* the extent of my freedom or even ultimately whether I have acted freely at all: I am almost entirely unable to judge the degree of freedom in any of my actions. Not only are we frequently wrong in our moral judgements of others, we are also so often mistaken in our judgements of our own actions. I can always imagine myself standing before some one who knows me better than I know myself, some one of great wisdom and authority. When I am feeling terribly ashamed of myself or proud of myself she[5] will point out the constraints under which I acted, the illness that I was suffering from, the congenital defects in my psyche, the poor examples and education that I had been given and I will end up understanding that I had, not diminished, but no responsibility for what I had done, that my behaviour had followed as night follows day. I can, of course, also always imagine her concurring with my sense of shame and rubbing it in, showing that my wrongdoing was even worse than I had imagined, opening my eyes to motives that I had subtly concealed from myself—self deception in all of us is rampant. In the case of deeds of which I was proud she could similarly show that there was no more cause for pride in them than in being tall or good looking or intelligent (and we all take totally ridiculous pride in these things or shame at our physical deficiencies) or she might say, "Well done, good and faithful servant". In my every action I can find reasons to show that I acted from necessity and equally in each and every conscious deed I cannot imagine having

---

4. "Can the whole of the universe be taken back to a *single* thing which, if it is appropriately specified, leads inevitably to elephants?" Peter Atkins, Creation, Chapter 1.
5. Could I be speaking about my wife? Or my mother? Or God?

no responsibility. I have no sureness whatsoever what the judgement would be (though I have to admit being on the side of Donne who ends one of his sermons saying that on the day of judgement unless Christ puts his hand on the scale holding our virtues "we shall be doomed, we shall be weighed down, we shall be swallowed up irreparably, irrevocably, irrecoverably, irremediably!"). How then can I possibly claim that I really do know that I have or do not have any freedom. An analogy from quantum theory might be helpful.

Although each individual event is random, the aggregate of events is certainly not. Although there might be difficulty in determining the motives of an individual action I have no doubt whatsoever that in my actions overall there is an element of free choice of good and an element of free choice of evil: although in any particular act I can imagine that it could be demonstrated that I had no freedom, no all knowing judge whatsoever or howsoever imagined will wholly exonerate or (I most sincerely hope) wholly blame the whole me. I am not here using some sort of ontological argument, begging the question of whether any such judge exists but showing that on balance I cannot believe myself not to have some element of responsibility for my actions and hence some element of true freedom.

If this is not convincing then in the end I think I will have to rely on the two assumptions in the opening lines of the Preface—that we do have certain knowledge and that the universe is reasonable and see where these assumptions lead. It is a little like any scientific theory: it can never be proved to be true but if falsified will have to be abandoned. If it fits in with other theories and appears to enhance our overall understanding then to the extent which and for as long as it has not been demonstrated to be false it will be accepted. If a satisfactory scientific explanation can be provided, then my philosophy would crumble. So far, I am on safe ground.

Nothing I have read and more importantly (since there are such huge gaps in my reading) nothing that introspection has revealed and nothing I have observed in others has led me seriously to doubt that for all its obscurity freedom of will really does exist and that its supposed emergence through the course of evolution fails to be convincing as eradicating its essential mystery. Even if—and I do not believe this—freedom could be limited to a mental state of acceptance or non acceptance of all that fate has to throw at us, the question of the nature of the being that could have such a mental state needs to be answered.

It is not the exercise of free will that is the present subject, but the finding of a starting point whereby we might examine whether there is anything in us that is beyond the ability of physics, and by extension evolution, to explain.

There is no correlation between the indeterminism of quantum theory and free will. The absorption or emission of a photon by an electron is subject to certain laws of probability. Physical, chemical and biological changes are mediated through such absorptions and emissions and are subject to these laws; evolution is the inevitable outcome given sufficient numbers of particles able to interact and sufficient time. Are we back to a clockwork universe, but with the clock replaced with the inexorable course of evolution? In the clockwork universe

God made the clock and set it in motion. In the evolutionary universe perhaps there is no need for a God at all? I see no essential difference in the causation of an orbit of an electron or an orbit of a planet, of the fall of an autumn leaf or the stalking of its prey by a tiger: all are subject to the universal laws of physics; given sufficient numbers and sufficient time nothing beyond these laws is required for an understanding. This is not necessarily an argument in favour of determinism for it does not rule out the possibility of genuine chance events. Where these laws themselves come from is another thing and there are still the problems of the eruption of matter and energy from nothing and, the subject of this chapter, whether human freedom demands something quite outside the laws of physics.

In my possibility of choice of working for another ten minutes or going to make myself a cup of tea and in the personal responsibility I have for my actions, there appears to be an inescapable something required within my own control which can influence the behaviour of photons and electrons. The problem is too weighty to contemplate just at the moment. I will go and make myself a cup of tea.

It might help to examine the process of making a choice: to examine what went through my mind when I decided to make a cup of tea twenty four hours ago, without for the moment defining what I mean by mind. Having read and altered what I had written the previous day and muddled through a further page or so over a period of almost two hours with no very clear idea of where I was going or even where I wanted to go, I was feeling a little weary and frustrated. My target is to write for two hours at the beginning of each day whether or not on any particular day I have anything at all to say. If there is plenty to write about the time goes very quickly and I am reluctant to finish on time, but there are other duties and under present circumstances I cannot afford the luxury of spending more time on what might well prove idle and self-indulgent speculation. Yesterday the going was hard and when I looked at my watch for the third or fourth time there were still five minutes to go. Jolly close to two hours, and a convenient point at which to finish. Allowing the temptation of finishing a little early could be put to good advantage by making a neat example of what I was trying to say: here was an instance where there did appear, however trivially, to be some element whereby I was free. I could choose whether to remain at my desk for another five minutes or to get up, stretch my legs and walk to the kitchen.

Could it have just been an instance of Buridan's ass but with the oats not quite equidistant and the impression of freedom just an illusion? To deny any element of freedom would ultimately be to deny any element of responsibility. It might well be that my character was just too weak to allow any possibility of resisting the temptation to get up a few minutes early but this might be because I had allowed it to grow weak over many years. Well, although there is obvious truth in this sort of sentiment, I do not think that this was the case yesterday. Whether I remained for a further five minutes or got up and left it was no great matter. I could easily have decided to remain seated and write another line or two—I know myself well enough to know that that was the case. It was a convenient point to break off and the two hours is not a rigid resolution. All right, grant that it was not a case of gross concupiscence. The facts of the matter were that I had reached a convenient stage in an argument (nothing that a sophisticated computer could not recognise) with insufficient time to start a fresh development as none was present and bursting to be expressed and a perfectly reasonable alternative came to mind. Given my character and lifelong habits what else could I have done but do what I did. A well programmed computer would have done the same. There, editing the previous jottings and writing this has taken me beyond the two hours: I must get up and leave it for another day. And yet I am free to stay here if I decide to do so . . . .

Let us leave it at that. Whatever simple or sophisticated arguments are used I still know that I have an element of real freedom in my actions and in any particular act I cannot determine the extent, if any, of this freedom.

To return to the central core: how could it be possible that man has some element in him that allows him at will and by an act of will to alter the physical state of photons and electrons within his brain? Is there some other mechanism acting in this instance so that the electrical impulses which are transmitted down the nervous system, resulting in physical action, are mediated by something outside the normal laws of physics?

I am aware, of course, that quite apart from the staggering complexity of the interconnectivity of brain cells of which there are apparently approximately one hundred thousand million, the cells themselves are comparatively huge and also of staggering complexity, having on their surfaces alone up to ten thousand filaments each consisting of four hundred and fifty amino acid molecules made up of around two hundred and fifty atoms . . . . and all in a seething maelstrom of activity and all are subject to the laws of quantum theory. The operation of these laws in living cells is still the subject of intense research and much speculation. I have no doubt that this complexity is necessary for consciousness to be possible but it seems to me that it cannot in itself start to explain it in so far as self awareness is concerned.

There is a temptation to speculate that in all living matter something beyond fundamental physics is needed but a moment's reflection on the infinitesimal gradation from inanimate to barely animate and barely animate to fully animate—from stones to viruses and viruses to bacteria, bacteria to amoeba and amoeba to tigers will make the speculation redundant.

Is there then anything in matter—animate or inanimate—which cannot be explained *even on its own terms* by science? It is, I think, generally accepted that physical theory, as expressed particularly in quantum theory, does not even on its own terms provide a sufficient explanation of the activities of any living cells, let alone advanced nervous systems. But then, even on its own terms quantum theory does not even explain itself. It is, I think, also generally accepted that there is no reason to think that physical science will not in time develop so that its laws will encompass living cells. There are still insights to be gained of the same magnitude as those of Newton or Einstein, Darwin or Schrödinger—there are still huge mountains to be climbed. I am not decrying the conquests (I have all my life spent and hope to continue to spend a large proportion of my mental energy in trying to come to grips with current scientific thought, to a much greater extent than current philosophic thought which I so often find sterile) but it seems to me that all the advances of science have served to deepen our understanding that there are immense depths to be understood and reinforces my belief that there are other avenues to explore, higher mountains still to climb. My expectation is that this will continue to be the case. The recognition of patterns of behaviour which is the subject of science, hugely complex though it is, is only one aspect of understanding. It might, of course, be more accurate to say that current physical theory does not even on its own terms provide a sufficient explanation of the nature of matter. If my belief is that living matter lies within the ambit of scientific understanding in the same way as inanimate matter, why should not man himself? The reason is simple. It is that within limits we can truly act in ways that are beyond the ability of science to predict, not because there are no reasons (as in the case of a quantum event) but because we freely choose so to act.

I am still groping. To try to move the argument on I will start with a theory and see where it leads.

Man does not have an "external" immaterial soul (a ghost in the machine) capable of interfering with the normal operations of matter but rather that the nature of matter itself is such that it can under certain circumstances be intimately united with an immaterial element. To put flesh on this thought (do I mean thought on this flesh?) so that it is not an empty form of words, that is the problem.

The need for the immaterial element, its nature and the nature of the partnership are themes which are now going to be explored but first let me make a distinction between the dichotomies of body and mind, and of body and soul by not treating mind and soul as synonymous.[6] By mind I mean a mental faculty as wholly explicable or inexplicable as any other matter. By soul, if such exists, I mean an immaterial element in man which informs all his actions, whether bodily or mental: if the theory is right there is no essential difference between bodily and mental.

Mind or intelligence, in the sense of mental agility, is a product of the complexity of the central nervous system and just part of any living organism and the dichotomy of body and mind in this sense is of the same order as body and strength. Soul is a non-material, and therefore non localised agent indivisi-

6. Even Aquinas, who has so greatly influenced Christian and particularly Catholic thinking, treats each as synonymous: "Relinquitur igitur, animum humanum, quae dicitur intellectus vel mens....." (It is therefore said, that the human soul which is called the intellect or the mind...) (S.T.I 75.2:Responsum)

bly present in every part of me.[7] The dichotomy of body and soul is more of the order of the painting in the unlit gallery and light. The soul suffuses the body, enabling it to become a person.

If I understood myself what I am trying to say[8] perhaps I could find the words to express my meaning, but I am still groping. The idea of soul is not some ad hoc add on because it is comforting to think that we are superior beings, with godlike qualities and immortal, (although these beliefs are doubtless my starting point) but a necessary ingredient in the attempt to get some glimmer of understanding of how we can have self aware consciousness, of how we can be truly responsible. In this search not only can we look beneath the surface—we must. Here is a reality, a thing in itself that does exist and that we can strive to understand. My primary experience is that I experience: I am acutely conscious that I am I and not you. (I was once inexpressibly hurt by a person whom I most deeply loved saying on the point of death, "I am so alone" when I was there all the time.)

What then must be the nature of this thing, soul? What is meant by non-material? What is meant by agent? What is meant by indivisibly present in every part of me? What is meant by me? Can this thing soul exist apart from me? Does it form the essence of me with my body, as it were, the add on? Is it, as is so often said, imprisoned within my body as in a cage and longing to be free?

In the end I think it does come down to personal belief or faith or disposition but we must not in examination of our beliefs distort observational facts or rationalise them using arguments that cannot be supported.

My views are largely the product of my upbringing but modified by trying to understand how it all fits together. Possibly in this writing my views will change, not I suspect radically, although they might, but more likely in rationalising bits and pieces of ill digested beliefs. I will start therefore with my current beliefs. Where these are matters of personal faith and not as it were rational discovery they will be placed in italics. They are the place I have reached, the platform from which I must journey on.

I have very little understanding of the nature of this thing, soul.

By non-material I mean not consisting of matter as discussed in Chapter 2: not made up of atoms or photons.

By agent I mean something that is active, something that can act on matter or possibly interact with it.

---

7. An idea most eloquently (as always) expressed by Augustine some seventeen hundred years ago: Creatura quoque spiritualis, sicut est anima, est quidam in corporis comparatione simplicior....... Nam ideo simplicior est corpore, quia non mole diffunditur per spatium loci, sed in unoquoque corpore, et in toto tota est, et in qualibet eius parte tota est.......(Also, a spiritual creation such as the soul, is in comparison with the body a simpler entity.....For therefore it is simpler than the body because it is not extended in bulk through a given space but in each individual body it is both wholly present in the whole and wholly present in any part of it you choose.)(De Trinitate VI 6)

8. It is all very well for Wittgenstein and his Wovon man nicht sprechen kann....

By indivisibly present in every part of me I mean that the soul is not to be looked upon as some as some tiny little spiritual substance located in the pineal gland or the cerebellum or anywhere else. As the fingers cannot be separated from the pianist (cut them off and he ceases to be one[9]*) neither can the soul be separated from any part of the body.*

Can the soul exist apart from the body? I think that the answer to this is that just as space-time-matter-energy is conserved so are all non-material substances: the soul cannot cease to be: the law of the conservation of spiritual substances.

I believe in the resurrection of the body.

How the soul could function apart from matter is another matter which might well be idle speculation. The question need not arise given a belief in the resurrection of the body in another space-time. There could be no possible meeting point between the two space-times, no more meaning in the question of how long the soul might have to wait before being united to another body than in how long God waited before He created time. The resurrection of the body is not only a matter of religious belief: it follows necessarily from this line of speculation (I nearly said reasoning). Will the matter be of the same nature as at present? If one is to believe Paul's insight or revelation the matter will not be the same: "Someone may ask, 'How are dead people raised and what sort of body do they have when they come back?' These are stupid questions. . . . The thing that you sow is not what is going to come, you sow a bare grain say of wheat or something like that, and then God gives it the sort of body that he has chosen. . . It is the same with the resurrection of the dead: the thing that is sown is perishable but what is raised is imperishable. . . . I tell you something that has been secret: that we are not all going to die, but we shall all be changed. We shall not all sleep, but we shall all be changed. This will be instantaneous, in the twinkling of an eye. . . ."[10]

The soul does not form the essence of me with my body some form of appendage. I am a souled body, a bodied soul.

The view that the soul is in any way constrained by the body is entirely false.[11] *Without the body it could not act at all. Of course I, body and soul, am limited: I am a creature. Of course there are those who long for an end to this vale of tears (if that is how they see it) and for the glorious life that is to come.*

Let us try to move forward a little.

Consider for a moment the implications of the belief that the soul is immortal and that after death will be united with another body possibly consisting of another form of matter. There is nothing greatly problemsome in this belief per se: if one form of space-time-matter could have been created capable of evolving

9. Joke

10. 1 Corinthians 15, 35 onwards The Jerusalem Bible, DLT 1974

11. What then of the saying of Christ, whom I believe to be the Son of God and God, incapable of error, in Matthew 26.41, which I believe to be substantially accurate: "The spirit indeed is willing but the flesh is weak." This is a statement on concupiscence of which more later, and not an identification of spirit and soul.

into structures which could be informed by immaterial substances, so presumably could another. The question that cannot be avoided is whether in any real sense it would be the same person, whether there would be a continuity of I-ness. There would not be much point in a different me going to heaven or hell. I know, without any shadow of doubt, that I am the same I as I have been ever since my earliest memories; I have had and have and will continue to have hopes and fears, joys and pains, can be rewarded and punished, but it is always the same I. Without this essential continuity all personality, all religious beliefs—all morality dare I say—become meaningless. How essential a part in this continuity does memory play, and is any memory possible without the physical structures in which memory is stored? The mechanism of memory is barely understood at the moment but huge advances are being made and there is every reason to believe that in the sense that we know how a computer's memory works, so we will learn how living creatures remember. Memory undoubtedly has a foundation in the physical structure of the brain: some permanent or semi-permanent change in the state of some neurons takes place as the memory is implanted, some in the conscious mind and others in the unconscious: the poem and the learning to walk. Our memories though unimaginably vast are only partly and fitfully accessible. Do I then, when I wake up for instance, remember that I am I? Or is it I who am doing the remembering? There are degrees of amnesia but total amnesia would equate with death if memory is taken to extend to all the learnt functions of the body. There do not appear to me to be sharp differences between learning to walk (of which I have no recollection, although this is not to say that the memory is not there somewhere) and learning to ski, between learning English and learning Greek[12]. Some activities require conscious striving and some do not but the end result is a learnt activity which then comes naturally and can be performed without thought. Is there also a smooth gradation backwards as it were from learning to walk to learning to digest food or are entirely different mechanisms at work? I suspect that there are no clear distinctions but that ultimately the term memory can be applied to the way in which an electron orbits or that quarks are organised within a proton.

And yet to some extent I truly have some choice in what I memorise (would that I had an equal choice in what I would forget!)

The notions of a disembodied soul or of one united with a new body do then present real problems in respect of memory and the only shred of an answer that I can think of is not wholly satisfactory—though perhaps not wholly unsatisfactory either. The I would be the same but with no trace of memory of this life—no trace of memory of friends or enemies[13]—but united again only on a much larger scale (encompassing all men) and all enraptured. This thought, at

12. Although later on it improved a little, learning in my case is rather a euphemism: I got 3% at O Level.

13. The thought that the new brain might have selected memories implanted into it seems utterly absurd.

first distressing, is worth dwelling on; it has echoes in the great Eastern religions and in much of the language of mystics of all times and all faiths.

It seems possible then that considerations of the immortality of the soul do not lead to inherent contradictions. My purpose is not to speculate about the nature of the after life or even to argue towards its existence at all but rather to explore whether the belief that the soul is immortal can be sustained. It might be well to start with a specific and carefully worked out historic statement of this belief which became the official teaching of Western Christianity for several hundred years and for Catholicism for a thousand years. Aquinas, following Aristotle, proposed a theory of substance such that any particular thing consisted of raw matter (prima materia, matter with no form at all) and form, an abstract notion. A tiger would be the union of the form tiger with raw matter. Matter was looked on as being sheer potency, stuff with ability to become this or that much as molten metal could be poured into this mould or that. Form was that which made one of the possibilities real, which actualised the potency. I think the modern equivalent would regard prima materia as the underlying stuff which can exist as energy or as an elementary particle but which can only actually exist as one or the other. The form is that which determines whether the matter is energy or this or that particle. These are the only true substances. All other things are merely arrangements (sometimes, as in the case of a tiger, exceptionally beautiful and complex arrangements) but still an arrangement and not an essentially new substance. Apart from man? Apart from man?

Full understanding of the properties of the particles will lead to full understanding of the tiger but not I believe to a full understanding of man.

Aquinas did not believe that the form tiger had any but an abstract existence but that the form of a man was his subsistent soul.[14] He reasoned (using an argument which I confess to not fully understanding) that the soul has to be subsistent because it is by virtue of the soul that man has an intellectual faculty which enables him to know all material objects and that whatever could know all material objects could have no material in it as this would impede or colour the knowledge of other material objects. "Therefore", he continues, "the intellectual principle which we call the mind or the intellect has an operation apart from the body. Now only that which subsists can have an operation per se. For nothing can operate but what is actual: wherefore a thing operates according as it is; for which reason we do not say that heat imparts heat, but that which is hot gives heat. We must conclude therefore that the human soul which is called the intellect or the mind is something incorporeal and subsistent." He then goes on

14. Respondeo dicendum, quod necesse est dicere, id, quod est principium intellectualis operationis, quod dicimus animam hominis, esse quoddam principium incorporeum, et subsistens. (ST I LXXV II Responsum) The 1920 Dominican translation has "I answer that, It must necessarily be allowed that the principle of intellectual operation which we call the soul, is a principle both incorporeal and subsistent." The New Shorter Oxford shows subsistent as being rare and giving the only relevant non obsolete meaning as "Existing substantially or really." Here, I think it means "Existing separately".

to argue that whatever is subsistent is incorruptible. He is very aware of the acute difficulties of understanding how a soul separated from the body can have knowledge or memory since the natural exercise of both of these is by way of material images[15] but comes to the conclusion that because there is an immaterial element in memory it will persist after death[16] and that knowledge can be gained by imprint of divine light[17].

It might be that these are answers to questions that are no longer asked or that greater understanding of the mode of dependence of knowledge and memory on the physical organisation of the brain mean that the answers are no longer wholly satisfactory. I have found that an attempt to understand the implications of the belief that we have immortal souls through an enquiry into the workings of free will means more to me than starting from the intellect. It might well be, of course, that the questions that have haunted me for so long are equally ones that are not of general interest—and I am still no nearer any understanding of the form of union of body and soul.

At some stage in the development of an embryo it becomes a human person[18]; some, spurred on largely by the abortion debate say at the moment of conception, some after twenty or twenty four or twenty eight weeks or some other more or less arbitrary time. I do not think in common sense, in general observation of the behaviour of mankind, a foetus is regarded as being a person. Many are spontaneously aborted without knowledge of their bearers; early miscarriages are not given funeral rights or treated with any dignity at all. I am not here holding a brief for the pro-abortion lobby, for in principle I am wholly opposed to abortion but I do think that language such as "Murder of an unborn child" is inappropriate and talk of its pain an exercise in something akin to the anthropomorphism used when speaking of the terror of the hunted fox. Of course events in the womb and in early childhood can have perverse or beneficial effects throughout our lives: "memories" in the sense discussed above are stored in the unconscious regions of the brain or other parts of the body but I have grave doubts that little babies or foxes actually really feel pain—are self consciously aware of pleasure or pain. I (surprisingly) have been a child and have no recollection whatever of the agony of being born or howling for my mother's milk. I have had children of my own and suffered agonies at their so obvious suffering. As a young child, not a baby, I remember (more than sixty years on I can still feel it) sharing in the suffering of a hen hanging upside down and being bled to death. Now, do not misunderstand me. I am not arguing that babies, born or unborn, do not have souls or even (although I feel it likely) that foetuses do not but

15. After a long and rather contorted discussion at the end of the response in I LXXXIX I he concludes " Si ergo patet, quod propter melius animae est, ut corpori uniatur, et intelligat per conversionem ad phantasmata: et tamen esse potest separata, et alium modum intelligendi habere.

16. Articulus V, loc sit.

17. Articuli II & III

18. This was also, I believe, the view of Aquinas, following Aristotle.

only to try to illustrate that we cannot infer the presence of a spiritual substance from the appearance of feeling pleasure or pain and conversely that we cannot equate the appearance of these feelings with true pleasure or pain which requires conscious self awareness.

To return to my central point. If we are to believe that we have immortal incorporeal souls then at some stage in the development of the embryo this thing soul needs to be introduced, externally as it were, and united to the embryo, to give it the capacity when sufficiently developed to think and to will. I am not trying to determine at what stage this happens as this would be idle, unnecessary and untestable speculation but rather to try to come to some clue to the understanding of the nature of the union—of the nature of matter and of the nature of the soul that could make such a union possible. I am not even trying to come to some definite statement that this is the way things are but only to find a possible solution. At the moment I cannot find even that. I can find not the slightest understanding either of the nature of the soul or of matter such that there can be a union between the two. And yet I do believe that we have immaterial incorporeal immortal souls as do and as have done the great majority of the human race from time immemorial.

I seem to have been writing in circles and I am not sure that I will get much further, so let me summarise by restating the two basic difficulties faced in denying the existence of a soul.

It is beyond my conception that any assemblage of electronic components could achieve one iota of self conscious awareness. Similarly it is beyond my comprehension that any assemblage of the fundamental particles of matter could achieve one iota of self conscious awareness.

Some explanation is needed for the possibility of the operation of free will. This would seem to require an interference with the apparently random nature of photon emissions or whatever other mechanism operates at neuron cell level in living creatures. These act in accordance with the laws of nature and lead to determinate (if unknowable) observable events in the creatures' behaviour, such as stalking a prey which is no less a matter of instinct than the migration of a swallow. In the case of man there appears to be an inescapable conclusion that the events at quantum level within the brain can to some extent be freely (although not of course consciously) self determined leading to an event in the world of observables which is neither fully determinate nor indeterminate in the normal sense. We can decide to walk in this direction rather than that just because we will it.

The only attempt at answers that I can offer at the moment to both questions are negative but not entirely without content. Perhaps they need another forty years of pondering. I would lay my bottom dollar on their not being falsified within this timescale.

The nature of matter must be such that no matter how far we progress in the understanding of its patterns of behaviour, further fields to be explored will always open up. At all stages we shall be as Newton, playing on the sea-shore, and diverting ourselves in now and then finding a smoother pebble or a prettier shell than ordinary, while the great ocean of truth lies all undiscovered before us. When the strings of a credible unified field theory are at last unravelled we shall not be at the end of physical enquiry; we shall be as far from knowing the mind of God as ever.

If this be error and upon me proved,
My writing's vain and God has never lived.[19]

With respect to understanding the nature of this non-material element in us, soul, an entirely different discipline to verbal ratiocination is needed to give substance to cold words. The name of the discipline is prayer and it in turn requires a purity and simplicity of heart that is the fruit of a lifetime's effort for the very few who persevere in this pursuit. This is easy enough to write, but is there any substance in it? Although it has been far from my lifetime's pursuit, I did for a time live with men for whom it was and (to the shame of my own life) I know that it is true.[20]

At the beginning of this chapter I said that it was being written in the first person because the arguments were from my own inner experience and I also referred to the two assumptions right at the start of the Introduction, that we have sure knowledge and that the world is reasonable. The Greeks had a saying "Nosce teipsum"[21]: "Know yourself" as being the font of knowledge. If the two assumptions are accepted they allow us to reject all notions of solipsism and to regard ourselves as typical members of the species, since there is no reason to think otherwise. It follows that I do have at least indirect access to your consciousness: that I know you have the same range of emotions and beliefs, fears and doubts that I have in all my various physical and mental states. I do not need to say, "I can only speak for myself": at least to the extent that my self knowledge is true I can speak for all mankind.

I cannot feel your headache or have direct access to your dreams but the more I understand the reasons for my own, the more will I understand yours and I imagine that this was the thought behind the inscription at Delphi.

It is a commonplace that self respect comes both from what we feel about ourselves and from what we learn that others feel about us. So here we have two almost platitudinous thoughts which are in conflict: that it is through self knowledge that I know you and that to know myself I need to know your opinion of me?[22] Rather, I think, they are part of an ongoing process. Apart from the suspi-

19. At least nearly as good a rhyme as Shakespeare's.
20. Two in particular, both Masters of Novices, one a Jesuit and one a Carthusian.
21. Odd that they should speak in Latin - γνωθη σεαυτον, I mean: it is inscribed on the temple of Apollo at Delphi.
22. cf Wittgenstein, almost at the end of his Philosophical Investigations: 'I can know what some one else is thinking, not what I am thinking. It is correct to say "I know what you are thinking", and wrong to say "I know what I am thinking." (A whole cloud of philosophy condensed into a drop of grammar.)' IIxi - Blackwell 1963 222e

cion mentioned at the beginning of this chapter that if I am really still I might be able to find knowledge in my innermost being, knowledge of myself and knowledge of you come from experience and observation. The only difference is that the majority of the experiences that I observe, that I am aware of, are doubly mine and not yours. Only I can feel my headache and I am aware of my own actions all the time and of yours only when we are together. Many years ago I started designing structures with very little previous experience, working at first on my own at home and two or three years later with one or two employees in an office but still largely in isolation, always with the suspicion that the methods of design were naive and possibly even unsound. Then we were given a contract to check other engineers' work (it is the duty of Local Authorities to check all structures before granting a Building Regulation approval and they may sub-let this work) and for a few years every new structure in a particular borough passed through my office. The methods used to overcome difficult problems, the assumptions made when it was impossible to get sufficient information, the thought processes and the forms of presentation were exactly the same as ours. There was, of course, some useful information gained but this was relatively minor; the gain in knowing that our modus operandi was sound was huge.

I suppose it is just possible that the same might be true of one's attempt to write a book of philosophy.

To return to the argument. The having of the headache is a purely personal experience. It increases our understanding of what a headache is but our reaction to it can only be learnt by self observation, by observing others who say they have headaches and by the reactions to ourselves of those with no aches. We can never know whether we are more stoical than others or more wimpish, but we can have a good guess. We have all met miserable people who we know consider themselves to be invariably cheerful and uncomplaining. Self knowledge and the knowledge of others in this sense is beset with uncertainties. Cromwell's famous phrase to the Scottish Kirk on the motives of the English in their advance into Scotland in 1650: "I beseech you in the bowels of Christ, think it possible you may be mistaken" can be applied to our own thoughts of our own motives as much as to our thoughts of your motives.

The chapter started with a poem, and I had meant to end it with one: "For Anne Gregory", by W B Yeats but there were problems with copyright. He expresses the thought, put into the mouth of a learned man, that only God could love her for herself and not for her beauty. It is readily available on the web.

CHAPTER 6

MORALITY: THE KNOWLEDGE OF GOOD AND EVIL

> **The woman said to the serpent, "God said, "You shall not eat of the fruit of the tree which is in the midst of the garden, neither shall you touch it lest you die". But the serpent said to the woman, "You will not die. For God knows when you eat of it your eyes will be opened, and you will be like God, knowing good from evil" . . . .When they ate the fruit their eyes were opened, and they knew that they were naked.**

When Adam and Eve transgressed they lost their innocence but they did not gain in knowing good from evil. Knowledge of what is good is a necessary precondition of being able to make a right choice in any action.

As we move through this book knowledge becomes progressively more difficult to attain. Knowing material things, so obvious at first sight, proved to be by no means so obvious; further layers of complications were added in our knowing living things and particularly other humans and ourselves. The very nature of abstract knowledge let alone its credit worthiness made for a wholly unsatisfactory chapter, and now one aspect of this, knowing good from evil, has at one and the same time a large measure of nearly universal agreement and an impossibility of ever knowing whether any particular act is good or bad.

When asked for advice by some one facing a difficult choice, a very common answer is

"You know what is right."

When not asked but you think you ought to voice an opinion in any case, you might well say,

"If I were you, I would think long and hard before deciding."

When asked but do not wish to get involved a third answer is also often heard,

"You will have to decide. I cannot make your choices for you."

This chapter will examine the processes of determining what is right—deciding on what is good or evil—from the first two of these three standpoints; the third is common to each of them and does not concern knowledge but personal responsibility. First of all there are a few brambles to be cleared.

Physical acts in themselves are morally neutral or, in the omne ens est bonum[1] sense, are good. The lightning that strikes you down is neither more nor less evil than the current of electricity in the person you murder by electrocution. The action of pushing when helping an old lady push her shopping trolley up a ramp is the same as that when pushing her over a cliff and the thrust of a surgeon's knife no different from that of a skilful assassin. Only intentions have moral values but actions which result from intentions (or sometimes omission of them) are called good or bad. No intention is morally neutral. Premature ejaculation may save the girl from rape but it does not lessen the guilt of the man. Fortuitous events sometimes prevent us doing things we had determined on and we are later so thankful that we have been stopped from doing something wicked and our virtue preserved. It is not our virtue that has been saved but our victim; we have been merely saved from public shame and punishment. We should be no less ashamed.

Although a number of my own views of what constitutes goodness together with my own prejudices and bigotry will inevitably intrude into this chapter, its prime aim is more to examine the processes of discernment of goodness than what is discerned and is practical rather than theoretical. How can we decide that this or that actual situation, action or choice is good?

Even though all men know what is meant by good it is not easy to define but it must be the business of this chapter to attempt to do so. A starting point will be to try to see where the knowledge comes from and to enquire into the extent of its certainty, its reality. If we do not know what goodness is how will we recognise it? How will we know it?

I imagine that *good* is one of the earliest and most often repeated of all of pleading and of praise. It can be a question, a statement or an instruction; it is all too often an irritable command. Later on it is a teaching to be obedient or quiet or kind, an instruction to be truthful or modest or prudent or brave or industrious. Disobedience is followed by punishment or threat of punishment, often in the distant future, and guilt complexes are laid down. As we grow up we form our own notions of what is good and bad but by and large these are only relatively minor modifications of what we have been brought up to believe. What I want to explore is the benchmark by which I judge my modifications. Why did Hucklebury Finn after much anguish decide not to do what he had been taught was right but tore up the note he had written saying where Jim, the runaway slave, could be found, even though he believed he would go to hell for doing

1. Every existing thing is good.

so?[2] Is this fictitious example a genuine human response or merely the sophisticated thought that its author put into the mouth of an unsophisticate?

Starting from the assumption that the universe is susceptible to rational enquiry, *good behaviour,* as a starting point, might be defined as that which is reasonable and evil that which is unreasonable. All conscious behaviour has some moral dimension; we cannot criticise any behaviour without forming some moral judgement although this may not be explicit. We cannot separate our opinion of the Government's policies from our opinion of their morality. Although we often do, we ought not to say, "It is not that it is morally wrong, it is just that I do not think that it will work." We might consider our child's request as unreasonable but would generally not think of it in terms of good or evil, but think how often a moral lecture is the response to a request. If a conscious act is trivial then the moral content is likely to be trivial, but not non-existent. Should I have my hair cut; do I dare to eat a peach? In our every deliberate action there are elements of doing our duty or being lazy, being kind or cruel, thoughtful or thoughtless, selfish or generous. When the mother praises her child for eating up she is laying down a moral norm: it is good to eat when we ought to eat; it is (we will learn later) good to ration our food to that which is sufficient and (later still) to abstain occasionally to learn self control. It is never indifferent.

Because the universe is reasonable moral theory is not capricious: what is right is also what is reasonable and there will therefore be a large measure of agreement between religions and between religious and humanist philosophies. Further, as we have evolved as a social species with a great deal in common with all animals and particularly mammals much of our morality is an adaptation, a refinement, a rationalisation of their instinctive behaviour. The faculty we have for discerning good from evil is called conscience. Conscience can be said to operate on at least three levels: it is a reasoned judgement of the relative moral merits of any case; it is the feeling of guilt (and less commonly of righteousness) following the performance of any act; it is an inborn light given to all men enabling them to distinguish good from evil (provided they are in command of their faculties).

Discernment can be gained by the constant application of the first of the three operations outlined above possibly by appealing to the last: it is unlikely to be furthered by the second. The third is the most difficult to access (some would say because it simply does not exist). The second is beyond our immediate control. The first is as complex as life itself and is a matter of continual judgement founded on inadequate premises. The subject is not without its difficulties!

---

2. " 'All right, then, I'll *go* to hell' - and tore it up. It was awful thoughts and awful words, but they was said. And I let them stay said; and never thought no more about reforming." Chapter 31.

A REASONED JUDGEMENT

For most of us most of the time our knowledge of good and evil is in accordance with what we have been taught as children: later formal or informal study has merely reinforced this teaching. By and large it does not occur to us to query the fundamentals. We accept that it is wrong to tell lies, to cheat, to steal, to injure or kill, to abuse children or those entrusted to us, to act with cruelty to man or beast, wantonly to break up another's marriage, to act solely in self interest without regard for others, to indulge in lecherous, treacherous villainy. Many of the precepts seem to be obviously right and not in need of any questioning but there are always situations where the application of the precepts are not so straightforward and give us pause for thought. I will later use the telling of a lie as a simple illustration, but any other vice could easily be treated in the same way.

If there are large areas of personal morality that seem to most of us to be reasonably straightforward – where by and large we know when we are doing wrong – there are also large areas where we are by no means so sure and the case is even more difficult when we are given the duty to legislate. It is wrong for me to kill in a private capacity but it might be my duty as a soldier. It might be my duty to send an army to fight. It might be my considered judgement that capital punishment is justified in some circumstances. Can a reasonable case ever be made for going to war or for capital punishment? These are stark choices but there are myriads where the moral element is less obvious and equal reason can be claimed on each side of an argument. There are matters such as monogamy or polygamy, abortion and euthanasia or assisted dying; there are political questions involving the extent of state control, of taxation, of business practices; there are the huge problems of the relationship of one nation to another including the need to preserve peace and stability at home one the one hand and balancing this with the common humanity of allowing refuge to those fleeing persecution or a share to those escaping starvation.

Often the arguments on each side are so powerful and the influence of narrow self-interest so difficult to assess that we simply do not *know* whether we are doing right or wrong. We can only form the best judgement we can. I say narrow self-interest because it does not take much reasoning to realise that every deliberate action must be in pursuit of self interest, of self satisfaction. (Even the satisfaction of being able to do something to prove this premise wrong!)

As an illustration of the difficulties that arise, it is worth looking at the morality of telling lies.

Lying is clearly wrong and yet there are times when it is generous to be untruthful (when taking the blame to protect somebody) or kind (when praising one's host's cooking) or courageous (when lightly agreeing to do something terrifying). Lies are a necessary part of all bargaining, of all diplomacy and of all politics. Yet, indisputably[3], it is wrong to lie and when we tell a lie we know that we are telling one as opposed to bargaining or being polite. Can a case be made for ends justifying means or is this too always wrong? We have the following dichotomies:

To lie is always wrong
There are times when lying is either good or necessary;
The end may never justify the means
There are times when the end may justify the means

If the first premise is accepted in each case, and these moral absolutes are very beguiling, then a way of reconciling the second is to look on them not as choices of good but rather as inescapable choices of the lesser of evils.[4] Let us look at a trivial example. Having consumed a truly horribly meal we can skip the first six prevarications and proceed directly to the lie direct,[5] " I really enjoyed that. It was delicious." Undoubtedly a lie, but a kind one and an act of politeness. What then are the choices of evil of which this is the lesser or least? The first must be silence which would be both cruel and ungrateful. The next would be to use some form of weasel words such as, "It is years since I had a meal like that" or, "Am I allowed to ask you for the recipe?" The only evils that I can see about these is that they are cowardly and could well invite the direct question, "Did you enjoy it, then?" to which the half truthful reply that the recipe was only wanted to ensure that it was never served in your own house would not be possible, and a further at least half lie demanded and on this analysis it would seem that the prevarication or half lie is probably better than the lie direct. There is also the problem of keeping up the first lie because if you own up to it having been a horrible experience once you are out of the house a report is sure to get back to your host sooner or later, and as a further evil nobody will believe a word you say if you make a practice of telling fulsome lies. On balance it is always to tell the truth when cornered but not to go out of the way to tell it when silence or simple prevarication are options.

Much the same sort of argument can be used when deciding on whether there are times when ends can ever justify means: there are situations when one's judgement may be that it is the lesser or least in a choice of evils—and whether this judgement in any particular case is right or wrong will not be known this side of the grave.

Lying is nearly always, almost invariably, wrong and in this respect there is a simple and fairly clear moral test: if the lie is self-seeking, self-serving or self-excusing it is ***always*** wrong.

---

3. Pronounced, indísputably and not the indísputably ugly indispútably. As in conscience so in language, we need to inform our æsthetic judgement and we do this by reading the great poets. Can one imagine Shakespeare saying, 'indispútably'?
4. It might be surmised that all that Jesuit training has not gone for nothing
5. cf As You Like It V.IV.70:

The other approach is to look at the uses of language. When you tell some one negotiating to buy your house, "We will not accept a penny less than one hundred thousand pounds" when in fact you would be absolutely delighted to get eighty five thousand, are you telling a lie? It is tempting to think that there is not a simple answer to this, that if the person you are negotiating with understands very well that you are negotiating and not putting your personal integrity on the line, that you are in effect saying, "we might be telling the truth or we might not; you will have to decide" then you are not lying. Against this, look at the codes of practice beginning to be enforced in the City where to make a misleading statement is a criminal offence. Can what is wrong for a corporation be right for an individual? It can be argued that the whole purpose of negotiation is persuade the other party that your minimum requirements are greater than they actually are. Is it then acceptable to deceive without actually telling a lie? An honest negotiator (which is not necessarily the same as a good negotiator) neither tells lies nor sets out to deceive. Like a good poker player he weighs up the odds and takes great care not to reveal his hand. That word *good*!

GUILT

We feel guilty when we are conscious of wrongdoing. It is a complex physical emotion which might have at least three sources: it might be a product of evolution, it might be a conditioned reflex or it might be due to contravening an inborn directly God-given sense of what is right. With respect to this third, it is both difficult not to invoke Ockham and in any case unnecessary since an holistic view of evolution can provide an alternative and simpler explanation at least in the sense (in the view of the sceptic) of requiring only one deus ex machina rather than two.

When a cat looks guilty when it has stolen the cream it might be a conditioned reflex because it has been schooled not to or it might be partly evolutionary in that it treats its adoptive family as its natural one with all its lore, laws and customs or a mixture of the two. With humans the great difficulty is in recognising which of the two is responsible, and in either case whether its morality is true or not. That everything which is prompted by evolutionary development is to be followed is at the root of the concept of Natural Law: there is no conflict between nature and morality. The world (that is, all that is the case, and this includes the facts of evolution) is reasonable but the reasonableness is not simple; we do not all have an identical genetic constitution: indeed, it seems very probable that each of us has a unique one and some of our genes will determine our inborn sense of morality. In general those genes that are responsible for long term survival of a species are thought to be more stable than others and genes that determine our inborn feelings of guilt will fall into this category so there is a large measure of natural agreement across mankind.

Guilt is of course not confined to moral matters. We can feel guilty about things which have absolutely minimal ethical content: the way we hold our knives and forks or parking in the Managing Director's place. My sister claims she still (in her late sixties) feels guilty if she uses a safety pin in her underwear: the nuns in her convent school said it was slovenly.

AN INBORN LIGHT

The still small voice is the presence of God but not as something superimposed but as part of our very nature and being. There is no element of mysticism in listening to one's conscience but rather being at one with ourselves, returning to our roots.

Here, it seems to me, is a possible complex answer to the paradox of our possessing a conscience which can truly guide us in the discernment of good and evil but being in need of being externally informed. The first strand lies in the difficulty of separating the unconditioned from the conditioned reflex. The second strand lies in the well nigh impossible task of distinguishing the various layers of natural desires, the promptings of "nature", and reaching the foundation: not bedrock but thousands of metres of very stiff clay above which are successively narrower strata of shifting sands. To find the bedrock we need to turn to religion but what religion? How are we to know that this particular religious belief is true, since there are many and they are contradictory? This is the subject for the next chapter but do not expect an easy resolution.

There is a school which says that all men have an inborn light enabling them to discern good from evil, that if we are really honest with ourselves we know what is right; I think most of us in our heart of hearts rightly or wrongly believe this to be true. I know I do, but there are at least two troublesome aspects about this belief. I view with the gravest suspicion claims of others to know what is right (and have great sympathy with Cromwell's "Always consider that ye might be mistaken") and secondly there have been periods in my own life where I have had to take hard decisions and the light has not been easy to discern—or perhaps it has but I have been unwilling to open my eyes to it. There is our own experience of St Paul's, "Now we see but in a glass, darkly" and the sentiment in Eliot's Little Gidding that deeds we once deemed virtuous are later seen to be far from it.

It is not the same as the light of reason, if by the latter is meant the ability to work things out for it would seem that this is largely a mechanical process and the result of evolution.[6] It is perhaps one aspect of knowing God in the stillness at the core of our being. It is a light that is always there and which can never be wrong but which can easily be missed or mistaken, a pinprick among

---

6. There is a line of argument as follows: If the only rational explanation of the rationality of the universe lies in the *created* nature of matter itself, then evolution could provide a sufficient explanation of the inborn 'light' of conscience. What is right is always what is reasonable.

the myriad of stars which alone can grow to a blazing light as we approach it or be lost among the innumerable host as we seek another which never gets closer or which might also shine brightly enough in our disordered imagination to obscure all else. But who is to decide what is ordered and what is disordered, what is healthy and what is sick? Can this be left to each of us individually?

Here there appears to be an irreconcilable paradox. If all men have an inborn light there is no need for an external authority. If there is need for an external authority, there is no need for an inborn light. The solution proposed by many religions is that some external authority, underwritten by God, is necessary to inform our consciences.

My own Church (the one Holy, Roman, Catholic and Apostolic Church) is especially authoritarian and its doctrine as laid down in the general councils and papal teachings is that we do have a light but by original sin concupiscence entered the world and the light in each of us was obscured and needs to be clarified by the teachings of Christ (God from God, light from light, true God from true God) who gave to the Church the authority to make plain His teachings. If this is the case (and it is my personal faith, my personal belief) then where is there a need for any inborn light in me? What function does it have? Does its sole judgement lie in accepting that my Church has the authority and in an abrogation of one's own thoughts. Does it clarify one's own thoughts allowing one better to discern what is truly good?

Forgive me if I indulge in a little introspection, but my own internal muddle is the only one to which I have more or less direct access. When I consider those aspects of my Church's doctrines that I find hardest to accept, or in some cases am unable or very nearly unable to accept (papal infallibility or the teachings on birth control for example) I follow my own conscience but accept, of course, that I might well be mistaken. I read and listen to what the Church says with as much care and reverence as I can but in the end follow my own conscience, reserve to myself my own judgement. In public and even in the privacy of my own thoughts I accept the authority of the Church, only sometimes to whisper inwardly, " e pur si muove". I would prefer to believe not in the personal infallibility of a pope or even in any particular segment of Christianity but rather in the developing beliefs of mainstream Christianity over the centuries. What I find truly amazing is that across two thousand years of splits and schisms and bitter dispute the general consensus has remained so constant. It sometimes takes a century or two to do so, or in some cases many centuries, but divisions do seem to be followed by healing and after the healing the same message is seen but more deeply understood. I see the Roman Church not as an immovable rock but rather a supertanker with the reforming movements as tugs. There are and have always been small cults with widely divergent views (irritating little speed boats heading mainly for the rocks); there have been some major differences as in the very sharp divisions in the 11th Century when the Orthodox and Roman Churches separated and again in the 15th Century at the time of the Reformation but for the great mass of Christians throughout the ages the beliefs and moral teachings have remained extraordinarily constant. There are remarkably similar phenomena in Judaism and in Islam although the latter, after the very early split, has remained very constant but there are great contemporary tensions.

None of this advances us much in the basic conundrum that if we have an inborn light[7], there is ultimately no need for an external authority and if there is need for an external authority, the inborn light could just as well be extinguished. A partial answer is given in the Papal Encyclical, Veritatis Splendor of 1993:

> *"It follows that the authority of the Church, when she pronounces on moral questions, in no way undermines the freedom of conscience of Christians. This is so not only because freedom of conscience is never freedom "from" the truth but always and only "in" the truth, but also because the Magisterium does not bring to the Christian conscience truths which are extraneous to it; rather it brings to light the truths which it ought already to possess, developing them from the starting point of the primordial act of faith."*[8]

There are situations in life where a choice between dreadful evils has to be made with the possibility of dire consequences to ourselves, our families and sometimes many other people. We are all faced with these choices to a greater or lesser extent in our private lives, in business and in public affairs. There are situations in business and in battle when we have to pretend to a confidence that we are far from feeling and make statements about which only the outcome will determine whether we have been telling the truth or not. No one, I think, would claim that when faced with a choice that entailed the balancing of complex issues a search into the innermost regions of our being would reveal the right course of action. This is obvious but each one of also knows that having made the choice we are often disturbed by the still small voice and, to say the least, suspect that we are doing wrong. If we know we are doing wrong we have a moral obligation to do otherwise; if we only suspect it, then we have to make what best judgement we can and get on with our lives.

Here we come to the first of the great paradoxes of this inner light. Although I do believe that it exists I also believe that I can never be absolutely sure that any action of mine whatsoever is good: I am never absolutely sure that I can discern it. If this is true how is it that we can inform our consciences? The an-

7. "This was, quite simply, a good person: you just knew it...From time to time one meets such people, people from whom goodness simply leaps. I may lack the theoretical foundations to explain it but the experience is too strong, too unmistakable to deny. These people have in common something extraordinary. What? Could it be God?, I asked myself. Is Christianity right? Is Islam right? Is God the explanation of human goodness? The answer struck me with unexpected clarity. There is no need to explain human goodness. It exists. It is a primitive fact….."Matthew Parris in The Times 15th March 2003

8. Paragraph 64

swer to this is quite complex but I think it lies in the fact that for most of us most of the time we are unable to achieve the simplicity that is needed and do not have sufficient confidence or courage to follow what we know to be right. Most religions say that we must inform our consciences by obedience to their teachings but some give the individual conscience primacy.[9]

This is the almost unanswerable case for the authoritarian church: since we are prone to error, an external authority underwritten by God is needed keep us true. But if this is the case, then conscience becomes not an inborn inner light present in all men, but a set of rules laid down by a man or a body of men—even if these are acting under God's instructions. In Judaism, Christianity and Islam the fundamental law is that received by Moses from God on Mount Sinai.

Is there a middle way? There is an inner light but the duty of a church is to help us to discern it by offering us guidance towards acceptance of what is right, guidance that should not necessarily be blindly followed but which should be treated with the greatest respect. All religions are authoritative to some extent and all are questioned by their members. Indeed, it is only through questioning that development can take place. Fundamentalist religions of all kinds strive to impose strict codes of belief on their adherents. As in so much of this book and my philosophy, there is no real contradiction but the case can neither be proved nor demonstrated. It has to be tried and lived and will be seen to be true.

Just as a handrail on a narrow bridge does not limit our freedom to cross it (in the desired direction)—indeed, if the bridge is very narrow and over a high gorge it makes the transit much more free and easy—so external moral guidance gives us the confidence to follow the light of our own consciences.

What if, after full and reverential consideration of our church's teaching, our conscience cannot accept it? For my own case (again, forgive me, but my own conscience is the only one I know, and that not too clearly—and yet I am sure, provided I am being honest, that what is true for me is true also for you though the circumstances might differ) a clear example of inner conflict came with the publication of the papal encyclical Humanae Vitae outlawing any form of artificial contraception years before this had any relevance to my own life. I had never been wholly convinced by the concept of Natural Law (nor for that matter wholly unconvinced) whereby some acts are deemed against nature and therefore intrinsically evil. It is the antithesis of Dryden's *"what Nature*

---

9. For instance, the Papal Emcyclical Veritatis Splendor. Paragraph 62 says: "Conscience, as the judgement of an act, is not exempt from the possibility of error.... Conscience *is not an infallible judge;* it can make mistakes.", and quotes from the Second Vatican Council, "Not infrequently conscience can be mistaken as a result of invincible ignorance, although it does not on that account forfeit its dignity; but this cannot be said when man shows little concern for seeking what is true and good, and conscience gradually becomes blind from being accustomed to sin." (Pastoral Constitution on the Church in the Modern World *Gaudium et Spes,*16)

prompted and no law denied"[10] but "unnatural practices" has become part of common speech. Whatever deliberately frustrates nature is seen as intrinsically wrong. The natural end of intercourse is conception. Ergo. As against this simple and beguiling doctrine are a host of other goods which might be equal, lesser or greater. Among these are the huge necessity to slow down the population explosion by means better than widespread but quite natural famine and disease; the regulation of the size of one's own family to that which is supportable; the ability for a woman to determine her own destiny; the mitigation in casual sex of avoiding an unwanted pregnancy; the control of the spread of AIDS; the fostering of mutual love and comfort in a couple without the fear of pregnancy—witness the countless millions of stable and devoted couples the world over. Then, there are the associated evils, greater or lesser. There is the removal from an appetite of any constraint, leading to a fragmentation of stable relationships and untold distress of abandoned partners and unsettled children; there is the huge commercial exploitation and the increase in materialism; there is selfishness and self indulgence and a lack of generosity and the trivialisation of the dignity of human life with all its accompanying impoverishment. It is a little like immigration. A generous and liberal nation will welcome strangers who wish to settle particularly if they are fleeing in terror of their lives; no nation in its right mind would allow unlimited immigration once this placed too great a strain upon its resources or started to cause serious social unrest. No absolutist position is tolerable in either extreme.

The avoidance of artificial methods of contraception might well be a counsel of perfection. It should not be an imposition unless it is clearly the will of God. We are free to follow our own wills if we will but we are not free to define good and evil: our freedom is only in the choice. My conscience is not clear in this matter. I can see no compelling reason for the Church's teaching and have a suspicion that it might be placing an unnecessary and harmful burden on its faithful but I am far from sure about this. Where my own private life is concerned there are my own legitimate appetites and desires; there are the wishes of my wife to take into account (though we may neither of us do wrong to please the other); where the wider community is concerned do I have a duty to preach dissidence if I believe the Church's teaching to be wrong at least in its assertion that it is speaking infallibly with an authority vested in it by Christ, the Son of God, God from God, Light from Light, true God from true God?

The recent advances in biological and in medical technology pose a multitude of ethical problems. They have altered in part the condition of humanity on which moral law is founded and as yet there is no consensus, religious or lay, as to what is right: we simply do not know. Ancient laws and rituals of cleanliness necessary for the safeguarding of life have been superseded by greater understanding of hygiene. Safe contraception has for the first time in the existence of mankind divorced the control of fertility from any but the most minor forms of discipline; this has led throughout a large part of the world to a majority of the

10. Absolem and Achitophel , 3

young living together before marriage (which might be a good thing); this and much greater longevity and the influence of a global media have led to a great weakening of the institution of marriage, to the great detriment of countless children and to the stability of society. Abortion and euthanasia remain items of passionate debate as do many activities in the biotech industry from genetic modification of seeds to experiments in the cloning of human cells. Apart from technological changes there are also those that stem from a greater (or sometimes greater and always fragmentary) understanding of our physical and psychological constitution; both the perception of the morality and the legality of many aspects of our lives have changed more over the last fifty years than in the previous thousand or even hundred thousand. Among these are the rights of homosexuals to live their lives and the rights of women to be treated on an equal footing with men, their rights to have abortions and to be priests, the general rights of workers and of children and of animals—the list is almost endless. Law makers (probably rightly) tend to err on the liberal side and it is a great pity that so many lawyers give an impression of confusing legality with morality. Whereas it is nearly always wrong to disobey the law it by no means follows that it is right to do what the law permits. There is at the moment no consensus between the authorities of the major religions and none within those of either Judaism or Christianity; the authority of all religions is very widely challenged by the young all over the world. I am not suggesting that the young are not moral but only that they are not at all convinced by the formal precepts of their religious leaders. That a consensus will emerge I have no doubt.

As individuals we have to do our best to reach an understanding of what is right and wrong, good and evil, to rid ourselves of bigotry and prejudice, to open our minds to truth and generosity and to lead as blameless lives as possible; as citizens we must perform a difficult balancing act in our choice of voting at all levels. If we do not find the exercise of our democratic rights difficult we are probably acting from prejudice.

It is not, to repeat, the purpose of this chapter to try to set out an ethical or moral code but to examine the means by which we come to one and what certainty we have that our knowledge of right and wrong is true knowledge. I started by looking at the belief that we have access to some inner and infallible light but this is clearly not tenable in any simplistic sense although it might be in all simplicity.

CHAPTER 7

RELIGIOUS BELIEF: THE KNOWLEDGE OF GOD

Be still and know that I am God[1]

In Chapter 5 I wrote,

> *Yet, when I am able to still my being sufficiently I do from time to time know of the existence of immensity . . . . If this experience is not universal I have a suspicion that the suspicion that such an experience might be there for the taking is universal.*

As a Roman Catholic I am unavoidably writing from this standpoint. Nevertheless I hope that most of what is written will apply to any member of any religion or none.

I was brought up a Catholic. Had I been brought up a Jew, given my inborn character, would I not now believe in Judaism? If I had come from a pious Muslim family would I not profess a love and belief in Islam? Or as a Hindu, Hinduism or a Sikh, Sikhism . . . ..? If my parents had been Humanists would I not proudly proclaim atheism and thank God that I was free of the frightful superstition of religion?
Although at their core most religions have a great deal in common there do appear to be major irreconcilable beliefs so what reasonable grounds do we have for professing this religion rather than that? Perhaps there is a hint of an answer in Wordsworth's

> *The child is father of the man*
> *And I could wish my days to be*
> *Bound each to each by natural piety.*[2]

1. Psalm 46
2. Concluding lines of "My heart leaps up..." and then used as a quotation at the head of Intimations of Immortality.

This might be a reasonable ground for remaining faithful to the religion we have been brought up in but it is not one for determining the truth of one's beliefs or of what we profess are our beliefs, for do we really truly deeply believe much of what we profess?

For example, I really truly deeply believe each of the articles in the Nicene Creed. Were I of almost any other religion I might assent to the first ten or twenty words[3] (although if I was a Buddhist I would have great reservations about the language and the use of the word God, but my core belief might not be all that different). Many Christians might have reservations about various articles in it but if I was not a Christian I would not even be agnostic about the remainder of the Creed: I would consider it to be wrong in fact. If I was an atheist I would believe even the first five words to be without content.

Since we cannot all be right, what reasonable grounds can any of us have for our own personal belief? Augustine agonises on the first page of his Confessions (I am sorry to put this in Latin but to put it in English is like translating Shakespeare and after all I am just following Wittgenstein's example in his quotation from the Confessions in the opening lines of the Philosophical Investigations)[4]:

> *Da mihi,domine, scire et intellegere, utrum sit prius invocare te an laudare te, et scire te prius sit an invocare te. Sed quis te invocat nesciens te? Aliud enim pro alio potest invocare nesciens. An potius invocaris, ut sciaris? Quomodo autem invocabunt, in quem non crediderunt? Aut quomodo credent sine praedicante?*[5]

Given a different preacher, given a different home background and schooling my beliefs would almost certainly be quite other than they are, yet I really do believe the faith I have been given. So does the believing Jew and the Muslim, the Hindu and the Sikh. We all believe that we are the recipients of a truth that God has revealed. Are we all, or all but one, mistaken? Is God malicious? We cannot all be right and God is for all mankind so terribly terribly hidden. All faiths have their miracle workers and although I am more inclined to give their wonders a little more credence

3. "I believe in one God, the almighty father, maker of heaven and earth, and of everything seen and unseen." Written in Greek in Nicæa in 325; widely promulgated through the Western Roman Empire in the Vulgate, St Jerome's Latin translation of the late 4th Century and still a question of debate as to the meaning of some of the original Greek.

4. A friend wrote "Yes, but you are not Wittgenstein."

5. Let me, Lord, know and understand whether appealing to you precedes praising you and whether knowing you precedes an appeal to you. But who, if he does not know you appeals to you? For in his ignorance instead of the one he may appeal to another. Or rather should you be appealed to in order that you may come to be known? But how will they appeal to one in whom they have not believed. Or how will they believe without a preacher?...

than I am sightings of UFO's I put them much in the same mental bracket—happenings which are probably susceptible to scientific rather than supernatural explanation.

What reasonable ground can a reasonable person have for his personal beliefs? (Remember, my first premise for the possibility of any philosophy is that the world is reasonable.) What reasonable God would have proclaimed His plan of salvation to a small warlike race who would so easily and so often betray Him, or through a rabble rousing Preacher to a small number of ignorant peasants who would pass on a message that would so often and so easily be distorted that after a few years and for the next two thousand there would be bitter dissent as to what the actual message was[6]?

What reasonable person can believe in this or that particular religion? Is it possible to believe in a God but in no particular religion? Is it possible to have layers of belief so that on a scale of nought to ten, belief that we exist would score ten (or at least nine point nine nine), God eight, Christ six, the Virgin Birth two and miracles at Lourdes about a half?

I repeat, I really truly deeply believe all the articles in the Nicene Creed. Am I being truthful? Some years ago in an effort (much as in this present writing) to clarify what I believed I wrote a long essay, a short book, on the Creed, taking it article by article and came to what I thought was a fairly orthodox Catholic belief and understanding to which I could subscribe without qualification. Of the two priests with whom I discussed it one said that there was no such thing as "fairly orthodox" and the other that he very much hoped I would not publish it. What my beliefs are are of no purpose to this present chapter which is about the nature of religious belief. I could of course argue the case for Christianity but this is neither the place nor the point.

Let me start at the beginning. There is no one who has not in early childhood heard of God; religion to a greater or lesser extent pervades all humanity and historically seems always to have done so.

But then all men, it is facilely argued, once believed that the earth was flat: they did at least have the evidence of their eyes. There appears to be precious little evidence for any religious belief. Could it have some evolutionary advantage? That there are genes that predispose us to be gullible seems likely; that there are ones that incline us to a religious belief is a possibility but I as far as I am aware there is very little evidence of the existence of any such. There is some evidence that stimulation of parts of the brain can be accompanied by religious feelings – the feeling that one is in the presence of something vast and mysterious. That gullible people should be more inclined to be religious than non-gullible does not contain in itself an explanation of where religious belief comes from. That there are people who seem more inclined to believe in the occult than others is

6. (My Lord, my Preacher, from the proud height of my twenty first century intellectualism let me become part of that rabble and rouse me.)

pretty obvious, and it does appear to be in their character rather than in the circumstances in which they have lived. Proselytising cult members are, I understand, good at picking out the type of person likely to be susceptible to their approaches and there is also a multitude of very intelligent men and women who are prey to ridiculous superstitions. I suspect none of us is entirely free of superstition although it is possible greatly to lessen the irrational sense of foreboding even if not to entirely eradicate it by deliberate contrary actions and the observation that no ills result. I suspect also that none of us is entirely free from intimations of immortality even if we regard this as a superstition.

A simpler explanation is that it is a human invention to make the task of governing more easy and indeed some political writings, ranging from Plato to Marx explicitly accept and promulgate this view.

Whereas the elaborate ritual laid down by Moses in Leviticus, Numbers and Deuteronomy can be read as a means of perpetuating the supremacy of the Levites, or many of the pronouncements of Rome as giving power to the College of Cardinals and clergy it is also true that nearly all the founders of major religions, or reformers of existing ones, are anti-establishment and by and large not out to make powerful or remunerative bases for themselves. Would that the same could be said of their followers. I do not think it would be possible to sway a person or a people unless there was already a predisposition to be swayed: to be effective all preaching has to be to those ripe for conversion. Could this predisposition be genetic? We are back to the previous paragraph.

Because the ability to reason is largely a mechanical process and itself the product of an evolutionary process it does not follow that either a belief in gods or bogeymen or a disposition to believe in them is genetically induced and yet I know that some people are more inclined to believe than others. We seem ultimately to be faced with a choice between two unarguable positions. Mankind has a universal belief in the supernatural as the result of evolution or alternatively he believes because the supernatural exists because this has been made known to him in some way or other.

It is, if modern geneticists are to be believed, quite possible that each separate species, including humans, can be traced back to one pair of ancestors. In our case we will call them Adam and Eve. If God made Himself known to them, then further explanation is unnecessary. That God has in some direct way revealed Himself at various times and to various people is a belief common to very many religions. By and large there is no need to doubt the sincerity of those who claim to have had the revelations, but perhaps they were suffering from delusions.

Is it possible to have a delusion that is other than a re-ordering of phantasms already in the memory? Could the idea of Deity be part of that re-ordering?We have all from time to time been filled with awe at the beauty and majesty of the earth, particularly when we are in the wilderness or mountains but also at dawn at sea when witnessing a storm there; we have been overwhelmed by the vastness of the heavens. Could belief in a God be merely an extension of these feelings? Could it be like a belief in Father Christmas which as a child we are reluctant to let go? It can be—and for most believers is—a convenient and comforting belief (clearly not understood by those who advertised on buses: *There is probably no God. Now stop worrying and enjoy your life.)* It can also be a terrifying one: *Thou heardst me truer than tongue confess Thy terror O Christ, O God . . . .*[7] And so from pleasure or fear we cling to a belief because there is no contrary evidence. Believers in an after life can never be proved wrong: if wrong, they are doomed to die in ignorance. The unbeliever cannot proved to be right, but if wrong he will indeed learn the error of his ways. Pascal summed up this thought in his wager: since there is no convincing proof of the truth of Christianity it is a better bet to accept it rather than be sceptical. "If you win you win all; if you lose, you lose nothing."[8]You are wagering an infinite gain against a negligible loss and under these circumstances it is sensible to take the wager.The great majority of us have had no mystical experience, no revelations, no serious delusions and yet the overwhelming majority of mankind throughout all time has had some sort of religious belief.*Be still and know that I am God*, sang the Psalmist of the God of Abraham, Isaac and Jacob, the God of Israel, of Rome and of Arabia, the one God worshipped by Jews, Christians and Muslims. And Lao Tzu several hundred years later and in a very different culture:

> *Without going outside his door, one understands everything that takes place under the sky; without looking out from his window, one sees the Tao of Heaven. The farther that one goes out the less he knows. Therefore the sages learnt without travelling; named things without seeing them; and accomplished their ends without acting.*[9]

There is no practical proof and no need for a wager. If we can still ourselves sufficiently and look for God we will surely find Him. The proof of the pudding is in the eating. *Oh taste and see that the Lord is good.*[10]

This is true for all men and at all times, but the stilling is not the work of a moment but rather the condition of complete simplicity which takes a lifetime's effort as Eliot says in the closing words of Little Gidding.

To say what we find when we find God, this is much more difficult because we become inarticulate although the attempts are remarkably similar in so many writings over the centuries and over the globe.

---

7. Hopkins, Wreck of the Deutschland
8. Pensées 3,233, pp 434-42 Léon Brunschvieg Edition 1914
9. Tao, Lao Tse XLVII adapted from Gutenburg
10. Psalm 34

I *know* there IS in which we shall be lost and found, emptied and fulfilled, utterly fulfilled beyond any imagining, utterly, wholly and completely enravished. This belief, this knowledge, is shared across continents and across time. In Western culture the tendency has been to try to analyse and articulate our understanding of the nature of this Being; in the East this very attempt has been thought to come between us and the All. Having been steeped in western tradition I have found myself through sixty years of pondering no nearer to a solution of the riddle at the end of the Introduction but applied to our knowledge of God.[11]

It is not reasonable that we should deny ourselves the use of our reason, that we should not stretch every nerve and fibre of our brain, that we should not strive with every particle of our intelligence, that we should not use all the learning possible, that we should not agonize throughout life to come to an appreciation, an understanding, a knowledge of Him whom we believe to be our creator, the source of all goodness, the one hope of all fulfilment and yet I know that it is in the stillness of my innermost being out of which all thoughts and cares have been emptied that I will find Him and not in all the learning in the world.

Truly, you are a puzzling God, my hidden God. I suppose there might be some sort of answer in the different times in which the activities take place. There is a time for learning and a time for contemplative prayer. Each should be daily activities and each within the circumstances of our lives. It is the duty of the professional theologian to spend the majority of his working day in research or teaching and the duty of the monk to spend the majority of his in prayer. It is the duty of society (that is to say, it is your duty and my duty) to decide who will be the theologian and who will be the monk although the exercise of these duties is a complex affair conducted through our daily lives and in the ballot box. The passion to know and the passion to satisfy idle curiosity can be difficult to distinguish but prayer is much more straightforward. It is a little like embarking on a long serious novel or poem (say The Brothers Karamazov or Paradise Lost): hard work and always easy to find an excuse for putting off starting but ultimately nearly always rewarding and occasionally enthralling.

To come back to earth. There is, it seems to me, a large element of consensus in a common belief in the existence of—I will call Him—God. There is no consensus whatsoever in the myriad of religions which worship Him in one form or other. What reason can I have for thinking that my belief as a Jew or a Hindu or a Methodist is true?

---

11. "I know that I sometimes do wrong and I am reasonable sure that when I am absolutely honest with myself I know the difference between right and wrong for me without having to think about it," with the gloss, "Here (and this is an aside) I come to an ambivalence I have never resolved: the thought that it would not be reasonable not to use every fibre of our intelligence in trying to come to any judgements and especially moral ones."

Faith, the little red Catholic Catechism says, is a supernatural gift of God, which enables us to believe without doubting whatever God has revealed. We know what God has revealed by the testimony, teaching, and authority of the Catholic Church.

The Church gets this authority from Jesus Christ. And Christ has all authority in Heaven and Earth, being the Son of God. I believe this to be true because I have the gift of faith. But for the supernatural gift this would be a transparently circular argument. How can one be rational about one's belief? It is, I think, not so much a matter of demonstrating, of proving, that one's beliefs are true but rather of coming to an understanding that they are not irrational. If true, they are eminently, indeed almost transparently reasonable; if false then the whole becomes merely a tissue of superstition.

Over the years I have come to an accommodation with what seem to me to be these powerful reasons for doubting that what one has been brought up to believe in is true. It consists in essence in a presumption that the religion in which one has been brought up is true and that other religions should only be looked to only if one's own proves impossible to believe. I do not believe that I, as a Roman Catholic, am alone in having a supernatural gift of faith, denied to a Muslim. But I believe that Jesus really lived, was the Son of God and God Incarnate; the Muslim believes that He really lived but was only a prophet, albeit a great one. I say "I believe in Jesus Christ, the only begotten Son of God"; in the Holy Koran it is written, "*Those who say, 'The Lord of Mercy has begotten a son,' preach a monstrous falsehood, at which the very heavens might crack, the earth break asunder, and the mountains crumble to dust.*"[12] Two supernatural gifts with on the face of it starkly contradictory beliefs. There is the possibility of a reconciliation in the thought that what Mohammed was decrying was not what Christians believe, that he was unaware of the mystery of the Incarnation: three separate persons in one indivisible God, and the second person not only fully the one God but also really man. Easy words to say; a lifetime's meditation to begin to begin to comprehend. It is not, I think, the Trinity that Mohammed is decrying but that the belief that the one God had a son who was not One with Him. There is an argument by Hans Kung that Mohammed should be viewed as the last of the Prophets. The argument can be developed as follows. Christians should accept that his testimony is God-given and should study the Koran in much the same way as they do the Old Testament: a revelation given to a particular set of people at a particular time; it does not particularly signify that in the former the period is historically after Christ.

12. Translation in Penguin Classics 1956, Mary, Page 37. c.f. also in Jonah, Page 69, and in The Cave: "...admonish those who say Allah has begotten a son.....a monstrous blasphemy is that which they utter." There are also three denials in Ornaments of Gold. and others in The Believers, The Night Journey, The Hordes, The Cow and The Table.

To repeat, the purpose of this chapter is not an exercise in comparative religion, an attempt to try to reconcile different beliefs but to see how one can reasonably say, "This is my belief". The answer lies in an understanding that we can only try to discern the will of God for ourselves, that for us to believe the faith with which we have been brought up must, if it is true, be His will (nothing can happen that is not) and that there are great depths in the truths behind the words we profess. The presumption must be that the beliefs we have been taught are true and we should treat them with reverence and humility and depart from them not out of idleness or fickleness or intellectual pride but only if they become impossible for us to believe. The resulting change (or the resulting affirmation) should lead to peace and joy and always action for the better. Steadfastness or conversion not accompanied by charity is but a clanging cymbal. I find in my own religion (as I am sure others do in theirs) very little that is not susceptible to rational explanation: almost nothing unreasonable. Although I see much that is reprehensible in its formal structures (as I am sure others do in theirs, for we, the members and rulers of the churches are human) and although I no longer accept all its teachings (and I admit I might be wrong in this) I have no doubt of its fundamental goodness. In it I find (as I am sure others do in theirs) a sure path to holiness from which I so sadly and continuously stray (as I am sure you do).

When I say I really believe, what do I mean? I do not know that there is a God in the same way as I know that New York is a city or in the same way as I know that Shakespeare (or whoever) was a very great playwright or in the same way as I know that I have a wife who loves me. I have for the most part a weak belief in a Being beyond my imagination. There are times when I feel that the whole thing is unreal and there are times when His presence is very real, but the latter are for me much rarer than the former. The knowledge that New York is a city is certain but trivial and certainly not worth dying for. That King Lear is a towering work of genius could conceivably be questioned by future generations and this could be their failure (much as I think it is mine that I am not moved by Dante) or it could be because Shakespeare's manifest faults were ignored because he was greatly in fashion in our generation. I am very sure of my judgement in this respect but can imagine a situation in which my Church in a papal decree should declare the reading of him or going to see his plays to be sinful and my being obedient to that decree. I am not that sure of my own judgement. As far as my wife is concerned, we have been married for many years and know each other inside and out. Yet every one knows of spouses who have been devastated on learning the true feelings of their partners so even thou They have not changed all that much over the last thirty years. I feel very sure of my judgement I cannot escape from the possibility that I might be mistaken. With respect to my belief in God I will not admit to any possibility of doubt even (I hope) at the cost of my life.

Instead of the simple catechism definition of faith as a supernatural gift, am I now proposing an act of will, or is this act of will itself a response of acceptance of the gift? There is certainly an act of will in the refusal to accept any gift and although in the acceptance of many there is only passive acquiescence there are also many that require a very deliberate response. Think only of some gifts of inheritance. Would a better catechism answer be that faith is a supernatural gift freely offered rather than bestowed? I suppose ultimately this is true but unnecessarily complex for a child.

All moral choices (the subject of the last chapter) are in the end the acceptance or refusal to accept God's gift of His love, and this is an act of untrammelled free will: it is our only true and complete freedom. It is a choice offered to every person and exercised in his following what he perceives to be right. In my case it lies in trying to deepen and purify the faith in which I was brought up.

I will believe because I am obstinate; I will believe because not to do so would involve loss of face; I will believe because I find hope and comfort in doing so; I will believe because I would be lost if I ceased to do so; I will believe because my conscience tells me to; I will believe (even though I do not) because I believe that it is the will of God that I should. I do believe because I have the God given gift of faith. My judgement is that there are elements of all the above in my belief.

My prayer used to be, "Let me not lose my faith but bring me to an ever increasing understanding and love of You, my eternal Truth"; since writing this it has become, "Guide me to the truth and an ever increasing love of You, my eternal Truth."

CHAPTER 8

A THEORY OF PERCEPTION: WYSIWIRT

At the instant of perception the object perceived is within us. Space and time vary for each observer; matter itself has an unimaginable wave-particle duality. Attempts to understand what things really are in themselves lead to hideous confusion and largely unintelligible language.

There is, I think, another approach although one which will not recommend itself to any one with a closed mind as to the possibility of there being a creator.

If we accept that the Theory of Evolution is at least broadly true then the great diversity of life has come about by minute steps over aeons, and the mechanism for each step is a chance genetic mutation which has survived in the struggle for survival. Some relatively simple forms have survived for millions of years but there has also been a steady increase in the neural complexity of other species culminating, so far, in man. By the very nature of the survival of countless chance mutations over thousands of millions of years increasingly complex forms are bound to arise and some of these to survive. Sooner or later, given enough time, intelligent life must emerge.

Whether the particular forms of intelligent life – this actual horse – is a matter of chance is not to be assumed. Chance events are the part of the essence of quantum physics, and they lead to statistical certainty. The greater the number of events, the greater is the certainty of outcome.

Before Mozart was conceived the Jupiter Symphony had only a potential existence in a rather lesser way even than the infinite number of possible statues which exist in a block of marble: not exactly no existence whatsoever but something pretty close to it. Once he had conceived it, it had a real existence theoretically accessible for as long as and to the extent to which it was in his memory. When it was written down it had a much more permanent existence, but still a potential one, as a manuscript is not a piece of music. (Although we can go into a music shop and say, 'I am looking for a piece of music by Scarlatti', meaning only the sheet music). When played and recorded on a compact disc, the disc has reduced the potentiality but it is still not a symphony. When the disc is played on a CD player the digital information stored on it is converted into the patterns of air compressions which make up sound waves. This might be music—it is not a point I am prepared to argue at the moment[1]—and even when it becomes a sound to a human listener it is not necessarily music (much of what others apparently find musical I find cacophonous). Only when heard by an appreciative and informed person does it fully become music as Mozart, its creator, intended it to be.

The whole of creation is some one's music. It only achieves its full reality in conscious perception.[2]

Berkeleyis quoted by Popper[3] as saying "Everything is surface...Their appearance is their reality." If he was not saying that there was *no* reality beneath an object as perceived but only that there was no *deeper* reality, then that I think is my position.

WYSIWIRT

The analogy between light and understanding is commonplace but not I think between light and consciousness. It is a little bit like asking what a painting looks like in the dark. There is clearly a difference of situation between a painting which cannot be seen because there is no light and one that cannot be seen because the viewer is blind. That the painting exists in the dark is undeniable but as far as we the viewers are concerned there is no picture. It needs light to be reflected from it to make it possible to be seen. It needs an eye of the right sort (and an informed one at that) if it is to be perceived as the artist intended it to be perceived. The lit and living world is as the painting in the dark; we are not blind but have eyes of the right sort—although they need to be informed—and consciousness acts as light. It is not that we need consciousness to be aware of the world but

---

1. Feynman, Lectures on Physics 1963, 38-6: "The problem has been raised: if a tree falls in a forest and there is nobody there to hear it, does it make a noise? A *real* tree falling in a *real* forest makes a sound, of course even if nobody is there. Even if no one is there to hear it, there are other traces left. The sound will shake some leaves, and if we were careful enough we might find somewhere that some thorn had rubbed against a leaf and made a tiny scratch that could not be explained unless we assumed that the leaf were vibrating. So in a certain sense we would have to admit that there is a sound made. We might ask: was there a *sensation* of sound? No, sensations have to do, presumably, with consciousness. And whether ants are conscious and whether there were ants in the forest, or whether the tree was conscious, we do not know. Let us leave the problem in that form."
2. Pursuing the analogy of the symphony it would follow that the closer the perceiver is to the mind of God the closer to full reality is his perception, which is the commonly stated belief of mystics of many persuasions.
3. Conjectures and Refutations 6.8 quoting Berkeley Principles of Human Knowledge 87 & 88 where neither sentence appears exactly as quoted.

rather that through an act of perception we enlighten it: we alter its state from one of obscurity to one of clarity. Consciousness acts on the universe like a powerful torch in an art gallery in which the lights have failed.

This form of analogy is only an attempt to bring about an understanding of the thought expressed above, that there is no underlying deeper material reality than the world of which we are aware. We are not as blind men given sight but as seeing men given torches.

If what we perceive is the reality then such questions arise as to whether the sun goes round the earth or the earth really goes round the sun. According to Relativity there is no meaning in the distinction. According to Copernicus, who had a flash of insight, the former is a deeper reality in that it reduces the motions of the planets from complexity to simplicity. The power of this observation reinforced by the extraordinary synthesis of Newton seduced the understanding of the civilised world so that every schoolchild now knows that the reality is that the earth goes round the sun.

Poppycock.[4]

The deepest, the fullest, the most complete reality is what we perceive each time we perceive it and as we are perceiving it: the sun rising in all its splendour in the east and setting in its glory in the west, the long grass rippling in the May breeze.

Not only WYSIWYG[5] now but also WYSIWIRT: what you see is what is really there. Theories operate by some one having an intuition, a flash of insight (often subsequently proved false) and then devising experiments both to test the validity of the theory and to try to prove it false. No theory can be proved to be true.

Unperceived, the tree exists as some form of mass-energy in this complex entity space-time. Unperceived, (for example when asleep) we exist as some form of energy suffused by a subsistent soul which as a separate entity is unperceivable, even by self: I cannot perceive your soul and I cannot perceive mine. There is a temptation to think that I, body and soul, perceive myself as a sentient self-aware material object but this is not the case. A more accurate expression would be to say that I, a sentient and self-aware person, body and soul together, can with my senses, external and internal, perceive my body and by observation learn to know myself. I perceive you in the same way but I am not directly aware of your inner senses (I cannot feel your headache or your terror) and I am not continu-

---

4. "Philistine ignorance of science is still, in some quarters, thought witty and clever. How else to explain the following little joke, by a recent editor of the London *Daily Telegraph*? The paper was reporting the dumbfounding fact that a third of the British population still believes that the sun goes round the earth. At this point the editor inserted a note in square brackets: [Doesn't it? Ed.]" Dawkins Unweaving the Rainbow, Allen Lane 1998

5. What you see is what you get: computer jargon accepted by my computer's spell check, and appearing in all authoritative modern dictionaries.

ally with you so my opportunity to observe you is doubly circumscribed. My soul has in itself the fullness of its incomplete being but the material element in me only reaches its fullness of being when observed by me or by any other intelligent being, as does yours. I can sometimes see in you beauty of which you are unaware and in these moments you are truly beautiful. This is most clearly demonstrated at those moments of highest intensity in our lives, when we are desperately in love and learn, after agonies of doubt, that the love is reciprocated. What was ugly and deficient in us in our own eyes is seen as beautiful and perfect in the eyes of the lover, and is truly so.

Metaphysics apart, it is a commonplace that we flourish or are crushed in others' opinions of us.

Here is a theory:

> ***Space-time-matter was created with the properties necessary to give it a potential to evolve into structures capable of receiving self-conscious awareness through the exercise of which the creation will be completed.***

This theory, that the universe was created ex nihilo by formation of wave-particles of energy and matter of a marvellous nature so that they would have an ability[6] to combine into ever more complex structures and in such numbers over such huge periods of time, that beings with forms of consciousness would surely evolve resulting (so far at least) in ones which are self-awarely conscious, enabling them in conscious acts of perception to actualise matter is not, of course, a scientific theory at all but it can to some extent at least be subjected to the same tests as any scientific theory.

If anything in it can be shown to be self-contradictory or at odds with observation or unquestionably redundant then it will have to be abandoned. If this is not the case it will remain a possibility; if it leads to an ever-greater integration of one's whole being over a long period of time as one ponders its implications it becomes a stronger possibility. I would not like to go further than this.[7]

---

6. An ability rather than a predisposition or tendency. Like Richard Dawkins (Unweaving the Rainbow pg 184) I too was captivated as an undergraduate by Teilard de Chardin and for all his romanticism still have a sneaking suspicion that he might have been right but can think of no sufficient reason for not invoking Ockham.

7. It will doubtless be objected that this is merely a cop-out, a deus ex machina, which does not actually advance one's understanding and one which has been implicitly or explicitly rejected by many philosophers, particularly those of the Enlightenment. This might be the case, but I have not seen it put as explicitly as above, and it has advanced my own understanding to at least the extent that the problems faced by Berkeley and Kant, for instance, have become much more real for me.

There will still of course remain two huge questions which were discussed in Chapter 5 with no satisfactory conclusion: what is the nature of matter that it can be awakened by conscious beings and what is the nature of the creature that can awaken this potentiality?

Before making any attempt to re-examine these questions I would like to return to the planets going round the sun. What then of a flash of insight such as Copernicus had? Do the planets really go round the sun, or does it really rise each morning and set in the evening. Is there any meaning in the question? For you and me it certainly rises and sets—we experience this nearly every day of our lives. The tracks of the stars and planets taking the Earth as a fixed centre is hideously complex. For centuries it was a test of learning to master Ptolemy's system, an arbitrary construction of layers of epicycles, adjusted so that the observed paths more or less matched the predictions. Given sufficient circles a very good match indeed can be achieved but the system becomes very unwieldy. Put the sun in the centre of the solar system and much of the complexity is removed. Add Newton's understanding of the elliptical paths and very precise forecasts become possible. Refine this by Einstein's two theories and predictions that are as close to current observable limits are achieved. Which is the more real description? The ultimate reality of the human experience of the sun rising in the morning and setting in the evening, with its unfolding pattern throughout the year and repeated each year has been made possible by the existence of a solar system—by the creation of matter which as part of its evolution would form a solar system.

Is there something inherently absurd about this idea? Look at the analogy of making a film remembering always that it is only an analogy. Is what appears on the screen an illusion or reality? In terms of the outside world it is of course an illusion: neither the characters nor the studio sets are real separately existing people or houses. In the context of the film that I go to see, while I am lost in it, it is real: that which the film makers intended to create has come into being. There is no reason why I should not be curious as to how the effect—I so nearly wrote illusion—was accomplished but here the analogy starts to break down. After thefilm analogy how about that of the conjuror? Are we, by science, trying to understand the grand illusion created by the Great Magician? The answer, I think is *No*. In the one case the conjuror sets out to deceive. In the other—well, I cannot enquire into the mind of God but I will proceed on the assumption that there is no intention to deceive just as I have the utmost confusion about what I mean by ultimate reality. Is it then the ultimate purpose of scientific enquiry to understand how the trick is done—to know the mind of God? The answer of course is that there is no trick, no illusion, and through science, certainly no attempt to know the mind of God. The task of science is to discover and unravel the marvellous webs of patterns of the behaviour of matter; the duty of mankind is to use this knowledge, which brings with it powers of prediction and manipulation, to the benefit of all mankind.

There is an alternative theory that matter evolved ex nihilo as a quantum variation of the void and that after this the accumulated interactions resulting from blind chance mutations over aeons is all that is required for the complex evolution of the universe. When speaking of the void care must be taken not to confuse this with empty space[8]: although neither can exist they are not the same thing and I find the ancient aphorism, *ex nihilo nihil fit* (out of nothing nothing can become) much more persuasive. It seems to me also that any attempt to use quantum physics as a sufficient explanation of the appearance of matter ex nihilo begs the question of the validity of the basic tenets of quantum theory itself, of Schrödinger's equation and the like, for which we have no explanation: were these too quantum fluctuations? More seriously, it is a fundamental requirement of science that any theory can be tested and this is patently impossible with any theory about the void. Any *scientific* speculation regarding it is nonsensical. Literally nonsensical: without content. Whether metaphysical speculation has any merit cannot so obviously be dismissed out of hand.

It is unquestionably true that in the light of the advance of science, religious dogma as to the nature of the universe has had to beat a continual retreat and there are very many who believe that the retreat will continue until God and the immaterial become altogether redundant. There are many who already believe that all notions of the supernatural are redundant but apart from the argument from extrapolation I have yet to see a remotely convincing case. Uncritical dogma is so often replaced by uncritical dogma: closed minds meet closed minds.[9] There are, of course, many religious scientists of all faiths and the questions of whether religious beliefs can be and ought to be subjected to critical examination in the scientific mode is the subject for another work.

There is another theory which I find much more beguiling. If one considers the case of a falling object, energy is conserved in the fall, the loss of potential energy being matched by the gain in kinetic energy. If we consider

---

8. PeterAtkins is explicit that at least as far as he is concerned this is not the case: "In the beginning there was nothing. Absolute void, not merely empty space. There was no space, nor was there time, for this was before time". Then he adds a seemingly redundant statement: "The universe was without form and void." The Creation, Freeman 1981 Page119. He also says in the conclusion of the penultimate section , "We can now collect these speculations together....We shall begin before the beginning, allow unbounded speculation...."

9. My mind is certainly closed to this extent: I regard all thoughts which tend to weaken my already weak belief in the existence of God as temptations to be resisted. It is open to the extent that I will honestly try not to blind myself to any theory but to look on all as dispassionately as I can to see whether there are any ineluctable stumbling blocks. So far there have not been any but the process has led both to development of religious belief and to deepening understandings of scientific theories.

the universe as a whole it is possible to regard its total energy, the combination of potential energy due to gravity on the one hand, and on the other, kinetic energy due to movement, heat energy and the energy bound up in matter and so on, as totalling zero by considering the potential energy as a negative quantity. If this is true (and I am not sure that I understand the concept at all) then ultimately matter has no substance whatsoever. There would still remain Descartes' Cogito: my consciousness cannot be reduced to nothing. What marvellous economy it would be in a Creator to fashion this huge and complex diversity of the material universe from nothing and ultimately consisting in nothing so that it could become real in the minds of conscious beings.

I am arguing that space-time-matter was created of such a nature that it could (*and given sufficient time inevitably would*) evolve into structures which would be capable of forming unions with, of being informed by, non-material[10] elements which by their conscious operations would complete their creation.

This still leaves the two huge questions: what is the nature of matter that it can be awakened by conscious beings and what is the nature of the being that can awaken this potentiality?

As far as the first is concerned I find it is like driving a car in a snowstorm: there is nothing to focus on. The effect of the irregular patterns of snow on the windscreen befuddles and mesmerises and so it is with this. The reason is that once again I am falling into the same old trap: trying to find out what things are in themselves. What they really are (how often do I have to remind myself) is what the we perceive them to be: there is no deeper reality, nor is it the task of scientific enquiry to look for one. Earlier this year I looked at the American Express building in Lower Manhattan from almost a mile away and could see perhaps a thousand tiny windows. I knew (not admittedly with absolute certainty since the building might have been evacuated) that inside there were thousands of men and women occupied in a multitude of tasks. Even if the building was empty, I knew (admittedly not with absolute academic certainty, but certainly with practical certainty) that inside it were hundreds of offices furnished with tables, chairs and computers which themselves are composed of thousands of electronic components.... Similarly when I look at my hand I know that the skin is composed of many layers, that each layer is made up of millions of cells, that each cell has a complex active structure of hundreds of thousands of components. Am I seriously suggesting that my visual perception of the outer shell of the American Express building is more real, of a deeper reality, than the bustling interior, that the appearance of the my hand is in some significant way more real than the complex structure within? Yes, absurd though it may seem at first sight(!) that is exactly

10. I would write "immaterial" but I want to avoid any suspicion of a pun.

what I am saying. For me, at the moment of perception, what I perceived was the external reality of a large building. This was no denial of the reality of the solidity of the structure or of the fact that it had contents, but these were no more real, of no deeper reality, that my view of the building. When Wordsworth looked at the majesty of London in the early morning sunlight of 31st July 1802 the clouds of photons emitted in that instant of time by the ever active sub-microscopic wave-particles that have flowed unceasingly through the ages like a slow river through the volume of space known as London, forming an ever changing assemblage of buildings – these particular photons were absorbed by optical cells in his retinas which then triggered responses in his optical cortex which in turn were relayed to various other parts of his brain and he became consciously aware of a sight which he thought beautiful. The vision was constructed within him.

The tables and chairs, as objects as we imagine them, only become such when perceived by us. To return to the opening paragraphs of this book, if we had a completely different nervous system they would have a completely different appearance. Given a sufficiently different nervous system a chair would have as different an appearance as sight does from smell. The nervous system we have is the product of evolution and so two diametrically opposed conclusions follow.

If evolution is entirely a matter of blind chance, then our sense perceptions have little to do with reality. What we perceive things to be is just an accident, and we are back to Aristotle. Worse, because if it is a matter of blind chance then his substances are also accidents!

If evolution is a matter of chance resulting in statistical certainty (if the universe is reasonable) there is a possibility that our sense perceptions are far from accidental, and it is this possibility that is being explored here.

Simple mechanistic theories such as Newton's (although he, of course, was acutely aware of their inadequacy) are quite simply wrong in their implicit assumptions of an external unchanging time and space. Relativity and Quantum Theory cannot in themselves either support or destroy any philosophy but in so far as they reveal an ever more shadowy picture of wave-particles which can only be described in terms of partial differential equations with a multiplicity of physical interpretations, they might be thought rather to support than to contradict the ideas developed above.

The nature of the being that can awaken this potentiality – man – now needs to be examined.

As I sat and typed this first draft on the morning of 14th March 2003, as in so many other parts of this most unrewarding and unequal task, I had no ideas at all. Six years later the position has hardly changed. The only thing to do is to force myself to write and see if any clarity emerges. Perhaps the problem is that I have a preconceived idea of the end that I am looking for, and the idea is simply false. The preconceived notion is that man has an immaterial imperishable individual soul, intimately united to the body, suffusing the body, informing the body (words fail) enabling the creature man to be self-consciously aware, allowing him to reason, and giving him freedom to choose between good and evil and conferring on him immortality.

Human consciousness is the most complex result of the evolution of ever increasing individual neural networks and many believe that self-consciousness is only a matter of degree. I have already said that I do not find this line of reasoning persuasive because there is no evidence of degrees of self-consciousness (unless it be in a human child's development) and because there does appear to be an immaterial element in it. If there is any content in the idea that matter generally is a complex potency, becoming actual only in conscious awareness, then the first glimmerings of an understanding arise. It has taken me most of my adult life of intermittent thought and insufficient reading to arrive this far. It will take at least some months and perhaps years of much greater and more coordinated effort either to advance or to come to the conclusion that it is a blind alley, a chimera, a failed experiment, but never a waste of time.

Finally, with respect to the consciousness of animals or other living or partly living creatures, I do not want to enter any speculation here. That they are conscious is undeniable, and matter presumably reaches a form of reality in their perceptions. Let us, as Feynman said, leave it at that.

A SUMMING UP: A CONCLUDING SPECULATION.

In the beginning there was a void: no matter, no energy, no space, no time, nothing.

There is always the eternally present One.

Matter and energy, anti-matter and anti-energy, four forms which taken as a whole are no more than the void, were created and marvellously fashioned so that vast numbers of particle-waves or wave-particles would emerge. By the same eternal and unchanging act, these are preserved in incomplete forms, from which over aeons living matter would evolve ultimately with the potential of union with non-material substances to produce self-aware intelligent beings which by acts of perception would complete the creation.

Against this speculation, many would say, is the fatal flaw that it cannot be subject to any form of experiment which could disprove it but this is of course equally true of its denial.

In favour it can be said that not only does it not conflict with the great scientific theories of the last two centuries, but that as the nature of time, space and matter and life are probed more and more, so simplistic "commonsense" views of their nature become untenable and materialistic atheism does not provide a satisfactory explanation. We have come full circle and Dr Johnson's *Thus I refute it* becomes not only the common sense. But also the truest one.

APPENDIX 1

AN INTRODUCTION TO MATHEMATICS

INTRODUCTION

This is written for those of us who have either never known or have completely forgotten whatever maths might have once been learned. It is a practical rather than a philosophic introduction and aims to provide only sufficient maths to be able to follow the arguments or examples given in the main text and the second appendix and it assumes virtually no knowledge.

For most of us, simple arithmetic is just that: we can count, add, subtract, multiply and divide without too much difficulty, and use these operations as part of our daily lives. We can manage decimal currency and most of us have a reasonable notion of decimals and of percentages but I will nevertheless dwell for a moment or two on the basic notations in common use and on the symbols that are universally used. It might be possible to use plain English to describe and prove all mathematical functions but this would require excellent visual perception and a high degree of concentration. Understanding (and indeed development) is greatly facilitated by the use of special symbols and diagrams.
The symbols which will be used are as follows:

1	One
2	Two
3	Three
4	Four
5	Five
6	Six
7	Seven
8	Eight
9	Nine
0	Nought or zero
∞	Infinity

+ Add or plus

- Subtract or minus

* Multiply. The asterisk is used rather than the more usual x to avoid confusion when the latter is used to denote some unknown quantity. When in algebraic expressions letters are multiplied it is usual to miss out the asterisk so that a*b is simply written ab This is also true in the multiplication of bracketed groups. Thus (a+b)*(c+d) becomes (a+b)(c+d) and x*(y+z) becomes x(y+z). Further, when strings of numbers are multiplied the asterisk is often replaced by the same symbol as the decimal point but its use is clear from the context.

/ Divide sometimes used horizontally between an upper and lower expression to indicate that the upper is to be divided by the lower

(. . . . .) Operations within brackets (which might be nested) are to be calculated first before any external operator is used. Thus 3*(1+2—(3/4+1/2)) comes to five and a quarter

[. . . .] Same as above but often used for greater clarity in nested brackets

{ . . . . } Same as above but very useful when an elongated bracket is needed

x A change in the value of some unknown or variable quantity x

δx A very small change in some quantity x

dx An infinitesimal change in some quantity x

∂x An infinitesimal with variables x, y, z change in only the variable specified of an expression . . . .

$\geq$ Greater or equal to

$\approx$ Approximately equal to

∇^2 A differential operator; this will be explained later.

DECIMALS

If asked to read '321.123' we would say "Three hundred and twenty one point one two three". There is an asymmetry here. To complete the symmetry after 'three hundred and twenty one we should have continued, "One tenth, two hundredths and three thousandths", because that is what the digits signify.

When writing numbers we use a decimal system or one that uses the ten basic symbols 0 1 2 3 4 5 6 7 8 and 9. For the numbers zero to nine they are used singly: "I have 5 cats". After this we move over one column to the left with the symbol 1 and use the ten symbols in their normal order in the first column to signify the numbers from ten to nineteen: "I wish I had 15 sisters". and the symbol 2 in the second column for the numbers from twenty to twenty nine and so on up to ninety nine when we start all over again with a 1 in the third column: "Here are 164 dollars". A table can easily be drawn:

Thousands	Hundreds	Tens	Units
			1
			2
			3
			:
			:
			9
		1	0
		1	1
		1	9
		2	0
		:	
		9	9
	1	0	0
	:		
1	0	0	0

It does not take much reflection to see that in each case the right hand number in each column counts in units, the second number from the right in tens, the third in hundreds and so on so that 35429 means nine plus twenty plus four hundred plus five thousand plus thirty thousand: thirty five thousand, four hundred and twenty nine. The right-handmost number is for the single symbols, the next for ten times these, the third for a hundred times these, or ten times ten, or ten squared or ten to the power of two (different ways of saying the same thing) and the third for ten time ten times ten or ten cubed and so on. 35429 could be expressed as $3x10^5 + 5x10^4 + 4x10^3 + 2x10^2 + 9x10^1$.

Note the series 10^4, 10^3, 10^2 . . . ...each reduce by one tenth and this continues: 10^1, 10^0, 10^{-1}, 10^{-2} . . . .10^{-6} . . . ., meaning ten, one, one tenth, one hundredth . . . ..one millionth . . . .and so on. Anything (except zero or infinity) to the power of nought equals one.

Using the normal decimal notation, a point is placed after the units column (the 10^0 column) with the first column to the right giving the number of tenths, the second the number of hundredths and so on. Any fraction can be converted to a decimal simply by dividing the top by the bottom: this is wasily tested using a simple calculator.

(There are a whole range of numbers that cannot be expressed either as any exact fraction (one with whole numbers on top and bottom) or as a recurring decimal. These are known as surds and examples are the square root of two or π.)[1]

---

1 If the square root of two could be expressed as a fraction $^a/_b$ where both a and b are integers, these could both be broken down to their prime factors. If this is then reduced

POWERS

A positive integral power is a shorthand way of expressing how many times a number is to be multiplied by itself: $x^3 = x{*}x{*}x$

Multiplying or dividing numbers that are expressed as powers is done by adding or subtracting the powers, as can be easily demonstrated by writing the numbers out in full: $3^2.3^4 = 3^7$ or (3.3).(3.3.3) = 3.3.3.3.3.3.3= 3^7. It follows from this that fractional powers stand for things like square roots or cube roots. The square root of nine is three. Thus the square root of three is written as $3^{1/2}$ whence $3^{1/2}.3^{1/2} = 3^1$ equals three which is as it ought to be. This is of course not true for a mixture of numbers. $3^2{*}\ 4^3$ can no more be multiplied to make an integer to the power of five than can two cats and three dogs make five animals of a particular species. Fractional powers greater than one need to be evaluated by expressing the power with the numerator an integer divided the denominator: for example $3^{1.5}$. 1.5 = 15/10 and $3^{1.5} = 3^{15}$ divided by 3^{10}.

A negative power is the reciprocal of the positive power: $x^{-2} = {}^1/_{x}{}^2$

A number raised to the power of zero equals one: $x^0 = 1$

The only other bits of maths needed for an understanding of the Special Theory of relativity are simple differentiation and the Binomial Theorem.

Pythagoras (in any right angled triangle the sum of the squares of the lengths of the two shorter sides containing the right angle equals the square of the length of the longest side—the hypotenuse) seems to crop up all over the place. It is easily proved but I do not intend to do so here.

Tensor calculus (vector and matrix calculus) is very helpful in some aspects of the Special Theory of Relativity but not essential; it is not easy to understand but it is essential for anything approaching a full understanding of the General Theory. Detailed discussion of the General Theory is beyond the scope of this work.

Before dealing with simple differentiation we will look at the binomial theorem: it is not too difficult.

To multiply algebraic quantities in brackets such as $(a + b){*}(c+d)$ we multiply each of the terms in turn: $a{*}c + a{*}d + b{*}c + b{*}d$, usually written the rather shorter form of

---

to its simplest terms and is not an integer, neither can be the square of the top divided by the square of the bottom. So all roots of all integers are either integers or surds.

For example say the answer is ${}^{1412}/_{1000}$. This becomes ${}^{22*353}/_{2*2*2*5*5*5}$ and cancelling all that can be cancelled gives an irreducible ${}^{353}/_{250}$. If this is irreducible so clearly is ${}^{353*353}/_{250*250}$ and the answer cannot exactly equal two. This argument can be used for any fraction whose numerator and denominator are both integers (whole numbers) and a general rule can be made: all roots of integers are either integers or surds.

ac + ad + bc + bd. That this gives the right answer can easily be tested by putting in numbers rather than the letters. The binomial expansion states that
$(1 + a)^n = 1 + na + n(n-1).a^2/2 + n(n-1).(n-2).a^3/(3{\cdot}2) + n(n-1)(n-2)((n-3).a^4/(4{\cdot}3{\cdot}2)$ . . . .where a stands for any positive or negative number, lying between -1 and 1, and n is the power to which (1+a) is to be raised.
Once again this can be tested by giving a and n some values and seeing what happens, say one and a fifth and three respectively. Then:
$(1 + 0.2)^3 = 1 + 3 \cdot 0.2 + [3{\cdot}2 \cdot 0.2^2/2] + [3{\cdot}2{\cdot}1 \cdot 0.2^3/6] + [3{\cdot}2{\cdot}1{\cdot}0$ . . . ...] This and all subsequent terms vanish as they are all multiplied by 0 somewhere in the middle).

Hence $(1 + 0.2)^3 = 1 + 0.6 + 0.12 + 0.008 = 1.728$ which is (of course) correct.

How about a square root—say the square root of four fifths? Let's have a go.

Four fifths is one less one fifth and the binomial expression becomes:
$(1—0.2)^{1/2} = 1 + [-\frac{1}{2} \cdot 0.2] + [\frac{1}{2} \cdot -\frac{1}{2} \cdot (-0.2)^2/2] + [\frac{1}{2} \cdot -\frac{1}{2} \cdot -\frac{3}{2} \cdot (-0.2)^3/6] +$ . . . ...

= 1 — 0.1 — 0.005 — 0.00025 — . . . ..
= approximately 0.89475.

A more exact answer is 0.894427191 and if the series is taken on for another couple of terms this will result. A more exact answer still is 0.8944271909999 but to reach this the series will be needed to be taken yet further.

There is no exact answer, only better and better approximations.
So far so good because we have largely been dealing with concepts which we have grown used to in everyday life. This is not the case with differential calculus which I suppose most of us have either never been taught, or have learnt without understanding or have completely forgotten but which is a basic tool in the whole of analytical science.

It is divided into two main branches, diffferentiation and integration, the one being broadly the reverse of the other.

DIFFERENTIAL AND INTEGRAL CALCULUS

Differentiation is a technique for calculating rates of change (for calculating differences in a changing function). Integration is technique for summing these differences.

The technique is used to determine the rates of change and summations of continually altering functions when simple addition and multiplication as in the case above is not possible. The standard text book example, and for many reasons probably the best one, is a body moving in a straight line. If it is moving with constant velocity calculations of this at any moment do not arise and computations of distances travelled over any period of time are straightforward

matters of multiplication and addition, just as in the hours of daylight. If the body is accelerating the calculations are not at all obvious. Ignoring air resistance, a falling stone will have a smooth acceleration towards the earth such that its velocity increases by thirty two feet per second (ten metres per second) every second. In order to be able to calculate what its theoretical velocity would be at any moment and how far it had travelled in
any given period of time a number of methods can be used including differential and integral calculus.
Differential calculus is used to determine precise rates of increase or decrease at any moment (the rate of difference or differential) and integral calculus to determine the totals of all increments.associated with such changes.

Imagine a ball being hit by a bat at an angle of approximately 45° to the horizontal (or to the vertical – it is the same thing) and a velocity of just under 60 m/s such that its horizontal velocity is 40 m/s and its vertical velocity is 39 m/s. A table could be drawn:

Time	Vertical Velocity	Height	Horizontal Velocity	Distance
s	m/s	m	m/s	m
0	39	2	25	0
0.25	37	11	25	6
0.5	34	20	25	13
0.75	32	28	25	19
1	29	36	25	25
1.25	27	43	25	31
1.5	24	49	25	38
1.75	22	55	25	44
2	19	60	25	50
2.25	17	65	25	56
2.5	15	69	25	63
2.75	12	72	25	69
3	10	75	25	75
3.25	7	77	25	81
3.5	5	78	25	88
3.75	2	79	25	94
4	0	80	25	100
4.25	-3	79	25	106
4.5	-5	78	25	113
4.75	-8	77	25	119
5	-10	75	25	125
5.25	-12	72	25	131
5.5	-15	68	25	138
5.75	-17	64	25	144
6	-20	60	25	150
6.25	-22	54	25	156
6.5	-25	48	25	163

6.75	-27	42	25	169
7	-30	35	25	175
7.25	-32	27	25	181
7.5	-35	19	25	188
7.75	-37	10	25	194
8	-39	0	25	200

Two things to notice about this table are that the horizontal velocity stays the same (air resistance has been ignored) whereas the vertical velocity shows the ball as rising to a maximum with its velocity decreasing at a constant rate, and then increasing again as the ball falls. The acceleration is that due to gravity, 9.8 m/s/s at first in the opposite direction to the ball's motion: that is, downwards. The height of the ball does not vary at a constant rate: in the first second it rises 34 m; in the fourth it only rises 5 m.

Differential calculus allows rates of change of velocity or distance to be calculated at any instant; integral calculus allows the computation of total distances travelled.

A sufficient number of simple differentials will be evaluated from first principles to enable an understanding to be gained and the results of other useful ones will be given but not derived as this can always be obtained from a maths textbook.

y = 6 There is never any change in the value of y and the differential is zero.

y =any constant. As above, y is always the same and its differential is zero.

y = x Whatever the value of x is, this is the value of y: it changes at exactly the same rate as x does and the differential has a value of one.

y = 2x If x changes from say three to six, a change y will at the same time change from six to twelve: the difference is double and the differential is two.

y = ax where a is a constant and x is a variable: the differential of y in respect of x is a.

y = 2x + 4 If x changes say from thee to six, a difference of three as before, y will change from ten to sixteen, a difference of six and its differential is still two.

y = ax + b where both a and b are constants has a differential of a.

Now for the sleight of hand.

$y = x^2$ When x changes from say three to six, a difference of three, y changes from nine to thirty six, a difference of twenty seven giving a differential rate of change of nine times. When x changes from three to four, a difference of one, y changes from nine to sixteen, a difference of seven and a differential rate of change of seven times—not nine times as before. *What differential calculus does is to calculate the rate at which y is changing with respect to the rate at which x is changing at any instant.*

To go back to the example, when x changes from three to three and a tenth (or three and ten hundredths), a difference of ten hundred hundredths, y changes from nine to nine and sixty one hundredths: a difference of sixty one hundredths and a differential change of marginally over six times. When x changes from three to three and a hundredth (a difference of one hundredth) y changes from nine to nine and six hundredths and one ten thousandth: the differential is now very close indeed to six.

As the increase in x gets less and less the differential gets closer and closer to six. When $x = 3$ the differential is 6: that is, 2*3. The differential of x^2 is 2x.

All the above is much more simply expressed using the specialised symbols of maths.

Dealing for the moment with two variables only, x and y, if one variable is defined in respect of another (for example $y = 3x^2—22x + 12$*) the differential of the one is defined as the value of a change in the one divided by the corresponding change in the other at any instant.*

y = any constant. That is, whatever value x might have y stays the same: any change in x implies no change in y and $\Delta y/\Delta x = 0/\Delta x = 0$

$y = ax + b$. When x changes to $x + \Delta x$, y changes from $ax + b$ to $a(x + \Delta x) + b$

$$= ax + a\,\Delta x + b$$

The change in x is $(x + \Delta x)—x$ which equals Δx

The change in y is $ax + a\,\Delta x + b—(ax + b)$ which equals $a\,\Delta x$

and

$\Delta y/\Delta x = a\,\Delta x/\Delta x = a$

$y = x^2$. When x changes to $x + \Delta x$, y becomes $(x + \Delta x)^2 = x^2 + 2x\,\Delta x + (\Delta x^2)$

The change in x is Δx

The change in y is $x^2 + 2x\,\Delta x + (\Delta x)^2—x^2$ which equals $2x\,\Delta x + (\Delta x)^2$ and

$$\Delta y/\Delta x = [2x\,\Delta x + (\Delta x)^2]/\Delta x$$
$$= 2x + \Delta x.$$

As Δx gets smaller and smaller, $\Delta y/\Delta x$ clearly gets closer and closer to 2x: it tends towards 2x and ultimately equals 2x when Δx is infinitely small: *When* $y = x^2$ *the instantaneous rate of change of y with respect to x equals 2x.*

It is given the symbol dy/dx

The rule for differentiating (for finding the differentials of) polynomials—sums of terms consisting of several powers of one or more variables—is as follows

$y = ax^n$ where a is a constant and n is any power of x

$dy/dx = nax^{(n-1)}$

For example, if $y = 3x^2 — 22x + 12$, this can be written

$y \quad = 3x^2 — 22x^1 + 12x^0$ and

$dy/dx = 3 \cdot 2x^1 — 22x^0 + 0$

$= 6x — 22$

The differentials of a whole range of functions can fairly simply be worked out from first principles in a similar way. The only other ones I will mention here are those of the trigonometrical functions of sine and cosine. In a right angle triangle the sine of the angle marked is a/c and the cosine is b/c

y = sin x	**y = cos x**
dy/dx = cos x	**dy/dx = -sin x**

Back to the falling stone. How far will it have fallen in say five seconds?

Its acceleration (a) at any instant is the rate of change of its velocity with respect to time (t). That is, acceleration is the differential of velocity with respect to time:

$a = dv/dt$. It is given as a constant of ten metres per second per second and

$a = 10$

We need to find a relationship of velocity to time which when differentiated will give

$dv/dt = 10$. This is a bit like doing anagrams: there are no simple rules but some anagrams are simple and sometimes a complicated one is seen immediately or can take hours to dawn, or patient trial and error to work out. In this instance the solution is simple:

$v = c + at$ where c is a constant.

The constant can be evaluated by putting $t = 0$ when we know that $v = 0$ and hence

$c = 0$. Therefore $v = at$.

So far, so good. Now, velocity at any instant is the rate of change of distance (m) with respect to time and to find a relationship of distance to time we need a function which when differentiated will give

$dm/dt = at$. Once again the solution is pretty straightforward:

$m = d + {}^1/_2at^2$ where d is another constant which in this instance can easily be shown to be zero so that

$m = {}^1/_2at^2$ and after five seconds the stone will have fallen

${}^1/_2 \cdot 10 \cdot 5^2$

$= 125$ metres.

This process is called *integration*. It is given the symbol $\int$

APPENDIX 2

RELATIVITY AND QUANTUM MECHANICS

SPECIAL RELATIVITY

Time

Imagine a vast level plain with a long straight railway track and standing on it a rigid train, a tube on wheels. Draw a line across the track at the beginning and end of the train and call the distance between these two lines L_0 metres, the length of the train at rest relative to the line.

I am now going to sketch two thought experiments, the first of which is simpler conceptually than the second but I am not sure that it could be carried out even in principle. The second is a bit more tortuous but will, I hope, remove any fuzzy elements.

i) Have two identical light activated clocks which are alternatively switched on and off as flashes of light of a certain frequency reach them. Place one beside the track at the front of the train and the other just beside it on the side of the train itself at the front. Label the position at the rear of the train 'A'. Place a flashing mechanism beside the track at A or on the side of the train itself at the rear, it does not matter. The clocks will measure the passing of time in seconds and to avoid any confusion readings of the clock on the track will be recorded as "t" seconds and those of the clock on the train as "T" seconds. Stop each of the clocks and set them to zero. Flash the light to start both the clocks at an identical time. (t = 0 and T = 0.) Take the train back several miles and then accelerate it until it reaches any steady speed you wish so that, if it had blacked out windows, as far as a passenger inside it was concerned there would be no way to know it was moving (except that he had felt it speeding up going backwards and then slowing down and the speeding up again but this time going forwards and he was not aware of any further deceleration). Stand a coin on edge on a table in the train and it will stay standing on edge, and one could play pool or table tennis with the balls behaving exactly as they would on tables on the ground beside the track. For all intents and purposes the train is an inertial frame.

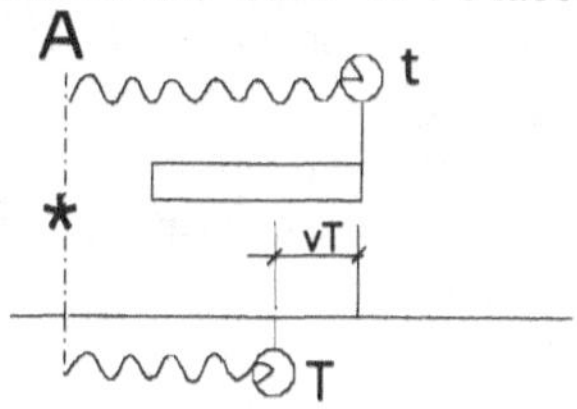

As the back of the train passes Position A the light mechanism is triggered and a light flashes at the rear of the train, or just beside the train at A: it makes no difference . When the light reaches each clock it stops it. The times elapsed between the light flashing and the clocks stopping will be Δt and ΔT on the track clock and the train clock respectively. (The symbol Δ means 'a change in.')

Now, and this is the crux of the argument, the cusp of understanding. By the time that the flash reaches the clock on the track the train will have moved forward and the light will still have some way to go before it reaches and stops the clock on the front of the train so one would expect this one to record a longer time. But as far as the clock on the train is concerned the flash of light (same flash, same stream of photons as those that travel along the line) has exactly the same distance to travel as the other: it is not aware that the train is moving and the clocks should therefore stop at the same time. The only way that this could happen is if the clock on the train runs more slowly than the one on the line as observed from the line. (*And vice versa*!) If the train is moving at v metres per second then when the clock on the track is stopped after a time Δt seconds has elapsed it will have moved vΔt metres and the light will take vΔt/c seconds to traverse this distance. Similarly when the clock on the train is stopped after a time ΔT seconds has passed it will have moved vΔT metres and the light will take vΔT/c seconds to traverse this distance.

Equating the times t seconds on the track clock and T seconds on the train clock then viewed from the line we get

$$\Delta t = \Delta T + v\Delta T/c$$

Similarly, viewed from the train we get

$$\Delta T = \Delta t—v\Delta t/c$$

and since the light is not aware that the train is moving,

$$\Delta t = \Delta T.$$

These are clearly consistent if $v = 0$ and equally clearly inconsistent if the train is moving.

Generally, if $a = b$ and

$a = b + c$ these can only be made consistent by introducing a factor n such that

$a = n(b + c)$

In this case we will introduce a factor R (the Relativity Factor) and evaluate it:

$\Delta t = R(\Delta T + v\Delta T/c)$ or $\Delta t = R\Delta T/c(c + v)$ ....(i)

$\Delta T = R(\Delta t—v\Delta t/c)$ or $\Delta T = R\Delta t/c(c—v)$ ....(ii)

Multiplying these two equations:

$\Delta t. \Delta T = R^2\Delta t. \Delta T(c + v)(c—v)/ c^2$ and dividing both sides by $\Delta t. \Delta T/R^2$ yields

$$1/R^2 = (c + v)(c—v)/c^2$$
$$= (c^2—v^2)/c^2$$
$$= (1 —v^2/c^2) \text{ and finally}$$
$$R = (1—v^2/c^2)^{-1/2} \quad \text{.................................(iii)}$$

Substituting this back into (i) above:

$\Delta t = R\Delta T = (c+v)/c = R(\Delta T + \Delta Xv/c^2)$ since $\Delta X = cT$ where ΔX is the change in distance in metres that the light flash has to travel because of the movement of the train in time ΔT. But although time is ticking on in the train, as far as a person in it, or the flash of light, is concerned the train is stationary, so $\Delta X = 0$ and $\Delta t = R\Delta T$.

i.e. $\Delta t = \Delta T(1\text{-v2/c2})^{-1/2}$

Similarly, $\Delta T = R\Delta t$.

i.e $\Delta T = \Delta t(1-v^2/c^2)^{-1/2}$

Since $(1-v^2/c^2)^{-1/2}$ is either equal or greater than one for all values of v, time in an inertial frame moving relative to an observer in an (of course) stationary frame is slowed by a factor $R = (1-v^2/c^2)^{-1/2}$ If the train is travelling at one hundred miles per hour or 100/60/60 miles per second, then over a period of three hours on the track clock (Δt = 3hr) the train clock will have moved $3/\{1—(100/60/60)^2/186000^2\}^{-1/2}$ hours

$\approx 3 \times 0.99999999999$ hours

or a difference much less than one millionth of one second so the practical effect at even relatively high terrestrial velocities is negligible. In a plane flying at a thousand miles per hour for three hours the difference is still less than one millionth of a second.

At a hundred and eighty six thousand miles per hour the times are different by a factor of only 0.999999961 or the moving clock will slow by approximately one second a year in relation to the stationary clock. The velocity of the earth in its orbit round the sun is of the order of 37000 miles per hour so it is easy to see that as far as most astronomical observations are concerned Newtonian mechanics are barely affected by relativistic modifications. At half the speed of light the clock moves at three quarters of the rate of the stationary one. As v approaches c, the traveller tends to become ageless as observed by the stay at home: at 185000 miles per second a traveller would age only five weeks to every year of the earth man.

Some fuzziness has crept into the argument presented above in so far as it might be argued that after the initial flash of light to start the two clocks, the one in the train is clearly not in an inertial frame for most of its journey: the train has to be accelerated in reverse, then stopped and accelerated forwards before it reaches its starting position. To ensure that all measurements take place in two inertial frames moving relative to each other, place the clocks at position A: clock t beside the track and clock T on the rear of the train. Place a mirror on the front of the train. Now, let the initial flash be triggered as the rear of the train passes position A at a steady speed v metres per second and the returning flash from the mirror stop each clock as it reaches it: only one and the same flash, remember.

The flash has further to go to reach the clock on the track (clock t) than the one on the train as the mirror has moved further on down the track by a distance $v\Delta t$ as before, and a further $v\Delta t$ on the return journey from the mirror and the three equations become

$2\Delta t = 2\Delta T + 2v\Delta T/c$

$2\Delta T = 2\Delta t — 2v\Delta t/c$

$2\Delta t = 2\Delta T$ which clearly reduce to the first set of equations and following exactly the same chain of reasoning as before we reach the conclusions:

$$\Delta t = \Delta T(1-v^2/c^2)^{-1/2} \text{ and by symmetry}$$
$$\Delta T = \Delta t(1-v^2/c^2)^{-1/2}.$$

We now have to confront the question of whether time really slows down in the moving object or whether it merely appears to the stationary observer: whether the two situations really are symmetrical. If the experiment was actually to be carried out and the train brought to a stop and the clocks were to be compared (remember, both clocks have been stopped by the flash of light, so they will not alter during the time taken to stop the train and we bring the clocks together for comparison) what will they read? Experiments have been carried out and the answer is that the train clock runs slow in relation to the track clock pretty well in agreement with the analysis above. The reason for this is not that the train really is moving in relation to the line but that it has undergone real accelerations and the track has not. These have altered its space time continuum (make something of that if you can). It is not that the clock slows down during periods of acceleration or deceleration and then reverts to its "normal timekeeping" but that being accelerated slows everything within it in relation to anything which is not being accelerated, microscopic and macroscopic, from a circling electron to the swing of a pendulum: slows time itself. As far as a person whose time is slowed is concerned he will not experience this but he will experience the acceleration: he will feel the force that is causing it.

This thought is unsatisfactory in that a "real" acceleration would seem to entail a lack of relativity; however, more of this later.

We have shown that $\Delta t = \Delta T(1-v^2/c^2)^{-1/2}$ where ΔT is an elapse of time measured on a clock in an object moving steadily with respect to another clock where the elapse of time is Δt. As $(1-v^2/c^2)^{-1/2}$ is always greater than one for any velocity in any direction, the moving clock will always appear to run more slowly than the stationary one, viewed from the stationary one. If one of the objects is really moving in relation to the other, its clock will really run slower.

We now need to show that $\Delta x = \Delta X(1-v^2/c^2)^{-1/2}$ and look at the meaning of this.

$\Delta x = \Delta X(1-v^2/c^2)^{-1/2}$ follows directly from $\Delta t = \Delta T(1-v^2/c^2)^{-1/2}$ by the straight substitution of $\Delta t = c\Delta x$ and $\Delta T = c\Delta X$ where Δx and ΔX are the distances that light will travel in times Δt and ΔT respectively. As with time, the consequence of this prediction is that a moving object is always shortened in the direction of its travel by a factor of $(1-v^2/c^2)^{1/2}$ in relation to the object from or towards or round which it is travelling. The question of whether it "really" shortens or just appears to from the point of view of the stationary observer has the same sort of answer as given for time dilation: it will really shorten with respect to the stationary object if it has experienced accelerations that the other one has not experienced. The position is not completely symmetrical with respect to time in that time will slow down in the moving object during its period of acceleration and will continue to run slowly as long as it is in motion. As it decelerates its clock will speed up again until it comes to rest when both clocks will tick at the same rate, but the accumulated loss of time in the moving object will not be regained. As far as length is concerned, the moving object will shorten during periods of acceleration, remain the same shorter length during periods of steady motion and lengthen again during periods of deceleration, until at rest both lengths are the same.

We now have two of the three fundamental equations of Special Relativity: $\mathbf{t_R = t_M(1—v^2/c^2)^{-1/2}}$ (where t_R and t_M are the times shown respectively by identical clocks in an inertial frame and in another one moving relative to it: relative to the stationary clock, the moving clock records a shorter time) and $\mathbf{x_R = x_M(1—v^2/c^2)^{-1/2}}$ (where x_R and x_M are the lengths of identical rods measured in a stationary frame and in one moving relative to it: relative to the stationary rod, the moving one will be shorter).

Mass

The suffixes $_R$ and $_M$ will be used respectively to denote a body at rest and one moving at a uniform velocity relative to it.

From the assumption to $m = m_R(1—v^2/c^2)^{-1/2}$ the corollary that $E = mc^2$ can be demonstrated quite easily:

$$m = m_R(1—v^2/c^2)^{-1/2}$$
$$m_R = m(1—v^2/c^2)^{1/2}$$
$$(m_R)^2 = m^2(1—v^2/c^2)$$
$$c^2m_R^2 = c^2m^2—m^2v^2$$

Differentiating with respect to t:

$$0 = 2mc^2dm/dt—2mvd(mv)/dt$$
$$c^2dm/dt = vd(mv)/dt$$

But mv^2 is an energy term and its derivative $vd(mv)/dt$ is a change in energy with respect to time so the above equation may be written

$$c^2dm/dt = dE/dt$$

and finally, integrating:

$$mc^2 = E$$

with the huge assumption that the constant of integration is zero: that the whole mass of a body can be converted to energy.

If the factor $(1\text{-}v^2/c^2)^{1/2}$ is expanded using the binomial theorem we get:

$$1 + v^2/2c^2 + 3v^4/8c^4 + 3v^6/48c^6 + \ldots$$

If v is small compared with c terms in v^4/c^4 and larger powers of v and c become very small indeed and the third equation can be approximated to:

$$m_M = m_R(1 + v^2/2c^2)$$ Multiplying through by c^2 and omitting the subscript $_M$:

$$mc^2 = m_Rc^2 + {}^1/_2 m_Rv^2$$

where each term has the units of energy, and where the third term is the normal expression for kinetic energy. The total energy of a body is made up of its rest mass and its kinetic energy, heat energy and so on. As c^2 stays constant, m must increase as v increases: this is in accordance with experimental evidence.

Conservation of total momentum

Finally in this brief exposition of Special Relativity is the derivation of a momentum equation which is relevant to quantum mechanics.

Assuming that total relativistic momentum is conserved in all inertial closed systems we get:

Total momentum (**P**) where **P** is a four dimensional vector is given by:

$\mathbf{P} = m_R c$ (That is, the total momentum of a body is the product of its rest mass and the velocity of light.)

$\mathbf{P}^2 = m_R^2 c^2$ But

$m = m_R(1 - v^2/c^2)^{-1/2}$

$m^2 = m_R^2/(1 - v^2/c^2)$

$= m_R^2 c^2 (c^2 - v^2)^{-1}$ or

$m^2(c^2 - v^2) = m_R^2 c^2$

$m^2c^2 - m^2v^2 = m_R^2 c^2$ Now

$E = m_R c^2$

$E^2 = m_R^2 c^4$ and

$m^2v^2 = p^2$ where p is ordinary three dimensional momentum. So

$m^2c^2 = m_R^2 c^2 + p^2$ and multiplying through by c^2

$m^2c^4 = m_R^2 c^4 + p^2c^2$ But

$E = m_M c^2$

$E^2 = m_M^2 c^4$ So finally

$E^2 = m_R^2 c^4 + p^2c^2$

This equivalence is used in the derivation of one of the basic formulae in quantum theory.

QUANTUM MECHANICS

Symbols

v	velocity in metres per second
a	acceleration in metres per second per second
m	mass in kilograms
p	momentum in kilogram metres per second
ν	frequency in cycles per second
λ	wavelength in metres
ω	frequency expressed as an angular measure where $\omega = 2\pi\nu$
k	wave number = $2\pi h/c$ (c is the velocity of light)
h	Planck's constant: the quantum of energy ($6.63*10^{-34}$ newton metre seconds or joule seconds)
ħ	$= h/2\pi$
V	potential energy measured in joules

E Energy measured in joules
ψ wave amplitude

Introduction

When a force is applied to a unconstrained particle it will cause it to accelerate in accordance with Newton's formulation: f = ma: force is mass times acceleration. Energy is needed for this to happen and is measured by the product of the force and the distance it has travelled: e = fx. Momentum is a measure of the mass of a body and its velocity: p = mv. Both energy and momentum are known to be conserved in any closed system.

Out of these readily understood notions arose Newtonian mechanics which allowed us to predict with great accuracy a huge range of actions from the behaviour of billiard balls to that of planets. It was found not to work when very small particles were observed (electrons and atoms) and in the first years of the twentieth century fundamental modifications to Newton's laws emerged. It is important to realise that quantum theory is not a new theory in opposition to Newton, but rather a reappraisal of him.

Photons, light particles, exhibit the character of being both particles and waves with corresponding frequencies and amplitudes. It is probably helpful to imagine both photons and electrons as particles: entities with clear locations and following clear paths, and it might be that this is the reality. It is tempting to think that they only exhibit wavelike properties when there are streams of them rather as molecules of water in a sea-wave, but this is demonstrably not the case. Single photons, single electrons, fired through two narrow slits arrive as discrete blips on a screen placed behind, but far from forming two distinct lines, as they would were they behaving just as particles, they form interference patterns as they are allowed to accumulate. If one of the slits is obscured, then a distinct but fuzzy line emerges. I am unaware if the spread of the fuzziness has been mapped in relation to the spread of the lines in the interference pattern which emerges in the two slit experiment.

All physical actions are the result of the absorption or emission of photons by electrons, which gain or lose energy in the process. This transfer is therefore in discrete packets or quanta, where the energy of a single photon is the product of Planck's constant and the frequency of the photon.

The fundamental wave equsation

De Broglie's equation, which applies to particles of all sizes, from electrons to planets, states that they exhibit wavelike properties where the wavelength is given by Planck's constant divided by the momentum of the particle:

$$\lambda = h/p$$

This relationship had already been derived with respect to electromagnetic radiation from the definition of Planck's constant and Maxwell's equation:

$$E = h\nu$$

$$E = pc$$

Combining these two equations gives p = hν/c and since the wave is travelling at velocity c

$$\nu = c/\lambda \quad \text{so}$$

$$p = h/\lambda$$

A couple of years after de Broglie suggested that all particles had wavelike properties, Schrödinger produced his awful looking equation for a particle of mass m:

$$-\frac{\hbar^2}{2m}\nabla^2\psi + V\psi = i\hbar\frac{\partial}{\partial t}\psi$$

A general wave equation, as explained in the main body of text (Page 32), would take the form

$A\nabla^2 h(x,t) + B\nabla h(x,t) + Ch(x,t) + D = 0$ where A, B, C and D are specific factors. It will be seen that Schrödinger's equation is of this form and is therefore that of a wave.

In order to make any sense of this at all we need to look at the equation term by term, to try to get some understanding of its physical significance.

First of all, the terms.

On the left hand side of the equation:

$\hbar^2$ (by definition) = $h^2/4\pi^2$ = $(6.63*10^{-34})^2/4\pi^2$ =$1.113*10^{-68}(Js)^2$ which when divided by 2m gives quantities of the order of 10^{-38} for an electron, 10^{-41} for a proton and 10^{-65} for a particle with a mass of one gram. (10^{-38} is one hundredth of one billionth of one billionth—a truly tiny number; 10^{-41} is one thousandth of this and as for 10^{-65}. . . . (It is because this quantity is so small—far below any experimental observation—that Newtonian physics works so well for ordinary observable objects.)

∇^2 is a shorthand form of writing "the sum of the second partial differentials. . ." (of ψ in this case). ψ is the amplitude, the height, of the wave and (here is the really difficult concept) its units are not in metres but in the square root of the probability of finding a particle of mass m in a particular location at time t (it varies both with distance and time: is a function of both space and time. It is not something that can be observed but is an abstract notion).

V is the potential energy of the particle in question.

On the right hand side:

i ħ is the square root of -1 "conditioning" ħ. i is a rather mysterious entity which has no obvious meaning in the world of real numbers but when used in equations can yield real answers. In particular equations such as (x +iy)(x-iy) = z it reduces to $x^2 + y^2 = z$ which is in the realm of real numbers. It is because we deal in real numbers in the observable world that it is $(i\hbar)^2$ that can generate a usable quantity rather than iħ itself. $\frac{\partial}{\partial t}\psi$ is the first partial differential of ψ with respect to t. It has to be a partial differential because generally ψ is a function both of space and time.

Schrödinger's equation was derived by combining Newtonian expressions for the total energy of a particle: $E = p^2/2m + V$ with Einstein's hypothesis: $E = \hbar\omega$ and de Broglie's theory that all particles have an associated wave function such that the momentum of the particle and its wavelength are related: $p = h/\lambda$. It is a fundamental postulate and its solutions have accurately predicted the behaviour of all matter within the tolerances of experimental error however much we have refined our methods of measurement. Newtonian mechanics failed both on the very large scale—plantetary motion needed relativity—nor on the very small where on the atomic scale it failed completely. The implications on the nature of matter which result from the undeniably accurate workings of ψ has been a matter of debate for the last seventy years and there is still no consensus—not that there being a consensus would necessarily mean that the truth had been found. There is an abundance of popular literature on the subject and one of the purposes of this essay is to enable non-scientists to read this literature with greater understanding. As with special relativity a selective (but it is hoped not eclectic) barebones of the theory is being presented with the aim of enabling the reader to follow the general mathematical argument.

Solutions to Schrödinger's equation, depending on the circumstances and boundary conditions in the space containing the particle sometimes yield continuous values of ψ^2: for example $e \propto \psi^2$ or $x \propto \psi^2$ (or the energy the position of a particle being only able to be determined as a probability—that is, statistically. In these cases we can only determine the probability of the particle being found at position x or having energy e. Sometimes it yields discrete quantities, often of the general form $X = nY$, where $n = 1, 2, 3, 4. . . .$and X might be, for example, the energy of the particle and Y its position. In this case the particle will only be found at these positions with corresponding energies. A good example is that of electrons orbiting atomic nuclei.

Rather as in the case of the Relativity, a rigorous analysis of even the simplest example of the practical workings of Schrödinger's equation is beyond the scope of this appendix but even an inadequate explanation of situation which is commonly found in text books should give some idea of the process. This is the case of a free particle in an infinitely deep well or trap, 2a metres wide, with impenetrable sides. By free is meant not subject to any external force or potential: it can still have energy in the form of, for example, heat. Imagine that the particle can move in either direction a distance x from the centre of the trap, pulled neither this way nor that until it reaches the sides when it comes into huge gravitational fields acting in the opposite direction to that in which it is travelling which prevent it from proceeding.

That is to say, when $\pm x$ is less than a, the potential energy V of the particle is zero and when greater than a it is infinite.

In this case Schrödinger's equation reduces to $(-\hbar^2/2m)\, d^2u/dx^2 = Eu$ where we replace $[\psi(x,t]^2$ with $[u(x)]^2$. That is, ψ which varies both in respect to position and time is replaces with u which only varies with respect to position.

This simplified version of Schrödinger's equation has solutions

$$u = A\sin(kx) + B\cos(kx) \qquad (1)$$

where A and B are constants and

$$k = (2mE/\hbar^2)^{1/2} \qquad (2)$$

When $x = \pm a$ the particle comes up against the barriers and $u = 0$. In this case

$$\pm A\sin(kx) + B\cos(kx) = 0.$$

And because of the boundary conditions either $A = 0$ and $B = n\pi/2a$ where $n = 1,3,5...$ or $B = 0$ and $A = n\pi/2a$ where $n = 2,4,6....$

Therefore $(2mE/\hbar^2)^{1/2} = n\pi/2a$ and $E = n^2\pi^2\hbar^2/8ma^2$

In Newtonian physics the energy level of a particle is continuous and can be zero (for a free particle at absolute zero, for example); in quantum physics the lowest possible energy for a free particle is $\pi^2\hbar^2/8ma^2$: it is quantised. For an electron with the width of the trap of the order of the size of a hydrogen atom (10^{-10}m) the base energy is approximately $1.1*10^{-67}/8*9*10^{-31}*10^{-20} = 1.4*10^{-18}$ joules or approximately 6 electron volts. (Electron volts are the units of energy used with respect to sub-atomic energy levels.)

As n has discrete integral values 1, 2, 3 . . . . the energy of the particle can only have discrete values and now we are dealing not with abstract probabilities but with but with something real and observable. The fundamental postulate of quantum theory

$E = n\hbar\omega$ is seen to follow from Schrödinger's equation at least in this situation (it does actually in all situations) and we have a formula for calculating the possible frequencies of the matter wave. We are still as far as ever from forming any imagination of the nature of the wave.

INDEX

www.ingramcontent.com/pod-product-compliance
Lightning Source LLC
LaVergne TN
LVHW090414210926
848480LV00002B/564
9780761851899